Fuel Cells and Alternative Fuel Systems

Topical Conference at the 2008 AIChE Annual Meeting

Philadelphia, Pennsylvania
16 - 21 November 2008

ISBN: 978-1-61567-208-0

Fuel Cells and Alternative Fuel Systems

Topical Conference at the 2008 AIChE Annual Meeting

Philadelphia, Pennsylvania
16 - 21 November 2008

ISBN: 978-1-61567-208-0

AIChE

Printed from e-media with permission by:

Curran Associates, Inc.
57 Morehouse Lane
Red Hook, NY 12571

Some format issues inherent in the e-media version may also appear in this print version.

Printed by Curran Associates, Inc. (2009)

For permission requests, please contact AIChE
at the address below.

AIChE
3 Park Avenue
New York, NY 10016-5991

Phone: (203) 702-7660
Fax: (203) 775-5177

www.aiche.org

Additional copies of this publication are available from:

Curran Associates, Inc.
57 Morehouse Lane
Red Hook, NY 12571 USA
Phone: 845-758-0400
Fax: 845-758-2634
Email: curran@proceedings.com
Web: www.proceedings.com

TABLE OF CONTENTS

Innovative Autothermal Reforming using Hydrogen Peroxide as Oxidant for Portable Fuel Cells

<u>In Hyuk Son</u>, Woo Cheol Shin, Sung Chul Lee, Jin Goo Ahn, Ju Yong Kim and In Seob Song

, Energy Lab, Samsung SDI, 575 Shin-dong, Yeongtong-gu, Suwon, 443731, South Korea

A compact and highly efficient processor for 200W portable polymer electrolyte membrane fuel cells (PEMFCs) has been developed at the Samsung Digital & Display Interface (SDI) Co. Ltd. Camping and caravanning, yachts and boats have high potential for penetrating the mark with the 200W range portable fuel cell. There is a high demand on the applications for electric power due to electric and electronic device however the sufficient power is getting increasingly difficult. To satisfy this demand, the reformer process should be compact, lightweight, fast start-up and high efficiency.

In order to support sustainable hydrogen economy, it is crucial to produce hydrogen cleanly and renewably. Ethanol is very attractive because of its relatively high hydrogen content, availability, non-toxicity, and storage and handling safety. More importantly, ethanol can be produced renewably by fermentation of biomass sources.

.

Ethanol reforming processes for hydrogen production can be generally classified into two groups: (1) steam reforming (SR) and (2) autothermal reforming(ATR). Compared with ATR, the SR process yields the higher hydrogen concentration in the product. However, for practical applications, the ATR processes are more attractive because they can be made more energy-efficient and the hardware can be smaller and lighter.

Air, which is usually used as an oxidant for ATR, contains a large amount of nitrogen, which can cause low the partial pressure of hydrogen, resulting in a decline in PEMFC performance. It is therefore expected that hydrogen peroxide (H_2O_2) is a suitable alternative nitrogen-free oxidant for ATR, providing concentrated hydrogen. In addition, H_2O_2 decomposition is highly exothermic (23.44 kcal/mole) for faster start-up and supplying heat source for ATR. Therefore, the ATR process using H_2O_2 requires no actuator to pre-mix the oxidant with the liquid fuel and no ignitor to start ethanol oxidation, resulting in higher energy density and efficiency reforming process.

For 200W portable fuel cell system, an series tubular type fuel processor consists of an autothermal reformer (ATR) and water–gas shift reactor (WGS), in which the overall integrated volume including insulation is typically small, 307cc (34cc ATR, 188cc WGS, and 85cc insulation). A hydrogen peroxide/ethanol solution (steam/carbon moral ratio = 2.95, carbon/oxygen moral ratio = 0.7) fed into the reactor by a liquid pump with low balance of plant consumed power (total < 2We). It starts instantaneously from room temperatures and reaches equilibrium within 3 minute. It produces hydrogen at 267 l h^{-1} (STP) with high concentrated hydrogen (62.5%) without nitrogen dilution in the process gas (62.5% H_2, 33% CO_2, ~3.5 CH_4, ~1% CO, dry gas base). This fuel processor of power density (~1,000 We l^{-1}), thermal efficiency (Maximum□82%, lower heating value) and specific power (Maximum 537 We kg^{-1}) are exceptionally high. We have operated the reactor for as long as 20 hours on the fuel process with no evidence of deactivation or carbon buildup. In addition, this fuel processor has demonstrated the ability to generate hydrogen from a number of hydrocarbon-based fuels such as ethanol, methanol, propane, sugar and bio diesel with hydrogen peroxide oxidant solution.

.

To improve the stability of hydrogen peroxide (H_2O_2) in the mixed solution, a stabilizer has been developed by Hansol chemical Co. Ltd. in Korea. By adding a stabilizer (< 300 ppm), the H_2O_2 can keep over 97.5% stability at 95°C for 1 hour, assigning over 99.9% stability for 1 year at ambient temperatures.

NON-THERMAL MICROPLASMA REFORMING OF HYDROCARBONS

Peter J. Lindner and R.S. Besser, Department of Chemical Engineering and Materials Science, Stevens Institute of Technology, Hoboken, NJ 07030
201-216-5257; rbesser@stevens.edu

Background

Environmental and efficiency problems of combustion energy production have made fuel cells extremely popular as a current topic of alternate energy research. The major limitation of hydrogen fuel cell systems is the lack of a readily available stream of hydrogen. Hydrogen requires large volumes due to its low energy density (11 kJ/L) and is highly combustible. Hydrocarbons are generally converted into a hydrogen rich stream through the reforming process. Traditional catalyst reforming processes have been used, but problems due to high device weights, catalyst poisoning issues, long start up time and excessive costs often make them impractical for commercial use [1-3]. Through the use of an efficient, non-thermal microplasma a hydrocarbon reforming system can be developed that circumvents current catalyst reforming issues.

The United States Army can potentially use microplasma technology to reform JP-8, a readily available military fuel and provide hydrogen to power fuel cells. This is beneficial as current portable power sources are supplied by battery power, which in large quantities can become too burdensome for the soldier. Furthermore the production demands to supply these batteries cannot always be met. An example of this was illustrated during Operation Iraqi Freedom in 2003 when the standard BA5590 battery (20 W, 154 W-hr) was in such short supply that six manufacturing plants working continuously could not meet demand [4]. Reforming JP-8 with microplasmas would develop a long-life, efficient, portable, reusable 'fuel processor-fuel cell' system, reducing the strain of battery technology in the military.

Introduction

The advantages of plasma reforming over catalyst reforming have motivated several research efforts in recent years. There have been various experiments published in literature using plasma as a reforming media for hydrocarbons like diesel, oil, gasoline, JP-8, methane and natural gas among others [1, 2, 5]. These investigations have shown the promise of plasma reforming, but have experienced issues achieving suitable system efficiencies. This project intends to show that microplasmas have advantages in comparison to catalyst, and offer overall improved system efficiency in comparison to conventionally-sized plasmas. This overview introduces the characteristics of plasma as a form of matter, the application of plasma as a reforming technology, and the theoretical benefits of non-thermal and micro-scale plasma operation.

Plasma

Plasma, the fourth state of matter, is defined as a neutral collection of ionized gas species and is the most abundant form of matter in the universe. Some examples include the sun and stars, the aurora borealis, lightning, fire and other such natural phenomena. There are also many every day examples of industrial and commercial plasmas. Items such as plasma TVs and fluorescent lights have found their way into homes and businesses. Industrial uses for plasma have been used for arc welders and plasma torches [6, 7]. Plasma technology has also been developed to etch and produce small, fine features prevalent in semiconductor fabrication.

Advantages of Plasmas Reforming over Catalysts Reforming

A chemical reaction occurs when a reactant converts to a product. The Arrhenius equation (Equation 1) governs the rate constant of a particular chemical reaction.

(1) **Arrhenius Equation** [8]: $$k = Ae^{-\frac{E_a}{RT}}$$

The dependence of the rate constant, k, is primarily due to the temperature and the activation energy of the reaction. Plasmas do not normally affect the overall activation energy of a chemical reaction, but rather they increase the temperature of reactive species and electrons reaching temperatures from 1 to 10eV (11,600 to 116,000K). The presence of high temperature electrons in the plasma diminishes the effect of the activation energy, and potentially offers a nearly instantaneous reaction rate. Catalysts are typically used to increase the rate constant via a reaction pathway with lower activation energy as shown in Figure 1.

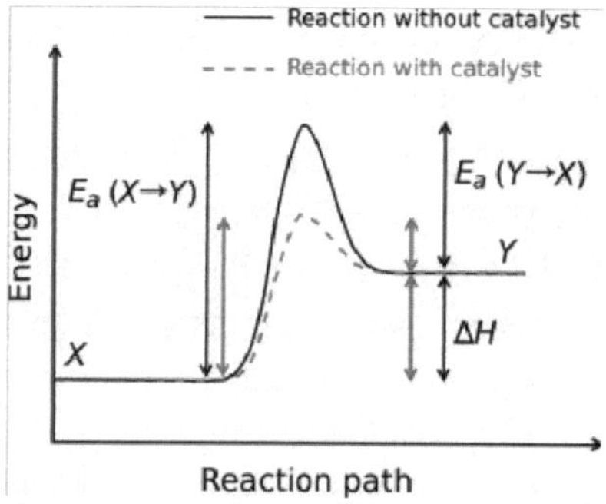

Figure 1. Reaction Path [9]

On the other hand, catalyst-based reforming processes have three major limitations; sensitivity to sulfur compounds, adverse effects of carbon deposits due to coking, and a long start-up time. Sulfur compounds cannot be tolerated at even minuscule concentrations as it acts as a poison blocking active catalyst sites. In particular, JP-8 fuel is known to contain sulfur levels up to 5000 ppmw [3, 10]. Catalytic reforming of JP-8 requires high operating temperatures exceeding 800°C, and would still be susceptible to poisoning from a monolayer of sulfur [3]. To avoid poisoning the catalytic reformer, the fuel requires a separate hydrodesulphurization (HDS) reaction. The HDS reaction converts sulfur compounds into hydrogen sulfide (H_2S). H_2S is then removed using a zinc oxide (ZnO) absorber. The addition of the HDS step not only complicates the system, but also necessitates an additional feed stream of hydrogen (Figure 2). An additional problem with catalysts is coking. Coking is the deposit of solid porous carbon or ash. Coke deposits result in a reduction of available active sites leading to overall catalyst deactivation. Carbon deposits are frequent when breaking down large carbon chains. Coking renders catalytic reforming ill-suited for hydrocarbon reforming processes. Furthermore, the start-up time for a catalytic reformer could take several minutes, rendering a power system useless during the start-up period [2].

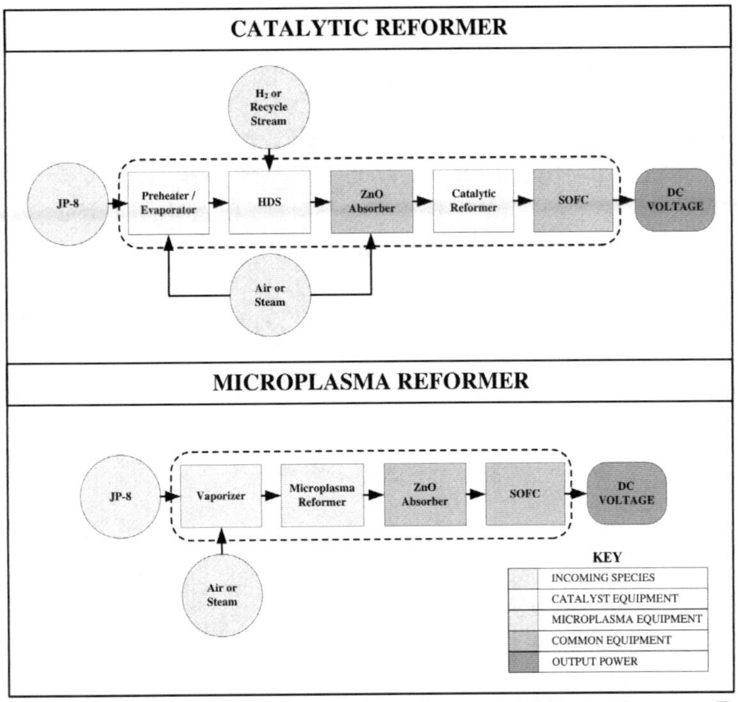

Figure 2. Comparison of Catalyst Reforming and Microplasma Reforming [11]

Plasma technology essentially avoids all of these issues. Earlier work demonstrated that plasmas can break down sulfur compounds present in hydrocarbon fuels, eliminating the need for the HDS step. This was observed in work at Stevens Institute of Technology in collaboration with FuelCell Energy Inc. of Danbury, CT [12, 13]. For this experiment NATO F-76 logistical fuel was sent through a non-thermal plasma device and a significant fraction of the organic sulfur components were converted into H_2S. Although coking is not expected to slow the reaction rate, the effect of coking on the microplasma device has not yet been investigated. Finally since the plasma is controlled through its own power supply, the start up of the device is instantaneous, determined only by the input power. For these reasons, we propose to further explore plasma technology, especially non-thermal, micro-scale plasma devices as a potential reforming mechanism for large hydrocarbon fuels.

Advantages of Non-thermal Plasmas

Thermal plasmas are defined as having electrons and the medium gas at the same temperature. Non-thermal plasmas occur under much lower electrical power, around only a few hundred watts. This process is defined as the electrons being at a much greater temperature while the surrounding medium remains at room temperature. Non-thermal plasmas do not provide significant energy to the system as thermal plasmas do, but they generate radicals. Non-thermal plasmas are ideal for chemical reactions due to their low temperature. They allow the energy consumption to be much lower, minimize electrode erosion and minimize the overall size and weight of the device [2].

Advantages of Microplasmas

Plasma technology therefore provides a non-equilibrium environment in which the generated, chemically-reactive free radicals promote chemical conversion and offer interesting prospects for hydrocarbon reforming [2]. The application of plasmas to reforming processes has recently been studied under various conditions but efficiency issues remain. At MITs Plasma

Science and Fusion Center (PSFC) a non-thermal plasma reactor called a plasmatron was tested with various fuels. Although very successful in chemical conversion, the device could not produce energy efficiencies beyond 74% [5]. Providing higher electron and free radical densities should theoretically increase the overall efficiency.

A plasma confined to at least one dimension 1 mm or less defines a microplasma [14]. As an emerging field of low temperature plasma science and technology microplasmas have much higher power densities (exceeding 1 kW/cm^3), higher electron densities (exceeding 10^{15} cm^{-3}) and increased surface-to-volume ratios when compared to conventional, large-scale plasma-chemical systems. A high surface-to-volume ratio imparts excellent thermal management and mixing characteristics that help maintain homogeneous, isothermal reacting volumes [15]. These microplasma characteristics present processing advantages for hydrocarbon reforming applications.

Moreover, conventional plasma processes generally operate at lower pressures allowing the system to meet the Paschen Minimum, minimum breakdown voltage. Table 1 shows the minimum breakdown voltage for various gases and the ideal electrode distance, d, for atmospheric pressure. Processing with microplasmas makes operating at atmospheric pressures possible due to the small distance between electrodes. Operating at atmospheric pressure minimizes equipment requirements and simplifies the overall operating system.

Table 1. Minimum Sparking Constants for Selected Gases [16]

Gas	Cathode	$V_{b, min}$ (Volts)	pd_{min} (10^{-3} Torr-m)	d_{min} @ 760 Torr (μm)
Ag	Fe	265	15.0	19.7
Air	-	360	5.7	7.5
CO_2	-	420	5.0	6.6
H_2	Pt	295	12.5	16.4
He	Fe	150	25.0	32.9
Hg	W	425	18.0	23.7
Hg	Fe	520	20.0	26.3
N_2	Fe	275	7.5	9.9
N_2O	-	418	5.0	6.6
O_2	Fe	450	7.0	9.2
SO_2	-	457	3.3	4.3

Microplasma technology offers many advantages in the devices design. They are generally small and can be designed to be any shape. The devices can easily include micro fluidic connections to allow for a flow through process. Integrated electrodes and heat exchangers for the individual microplasma devices are possible. Scaling up from experimental setup to the final design is simple as the reactors would have the same dimensions. Moreover micro fabrication devices are relatively inexpensive to produce. These advantages make microplasmas ideal for a portable hydrocarbon fuel reforming process.

Research Accomplishments

To date microplasma reactor chips have been designed and fabricated having different channel geometries using silicon microfabrication technology at the Cornell Nanofabrication Facility of Cornell University in Ithaca, New York. The microplasma reactors produced are microhollow cathode (MHC) devices with a variety of geometries.

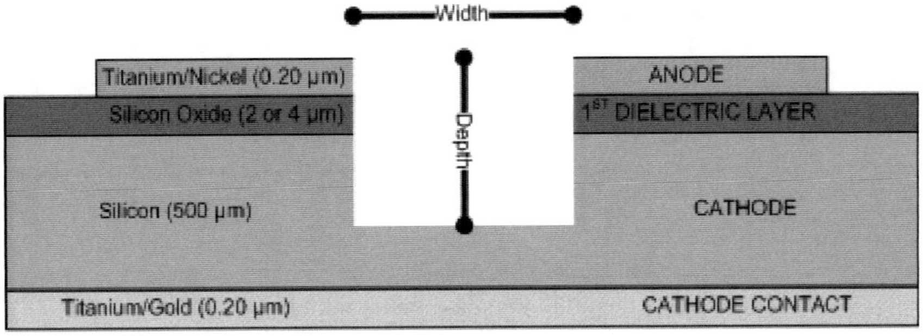

Width	25µm, 100µm, 250µm
Depth	25µm, 100µm, 250µm
Silicon Oxide	2µm, 4µm
Enhanced Perimeter	Yes or No

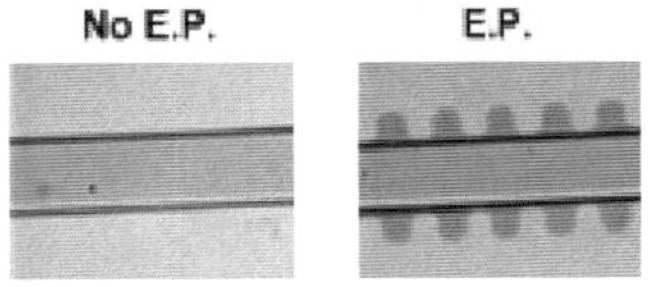

Figure 3. Microhollow Cathode Devices

After the fabrication of the microplasma reactors, a vacuum chamber was assembled to allow for testing and verification of plasma behavior. Initially tests were done with various inert gases such as helium, neon and nitrogen. To date inert plasma experiments have shown our devices to be comparable to literature. We have also observed the effects of device geometry and the degradation of the devices. Experiments were tested using the setup shown in Figure 4.

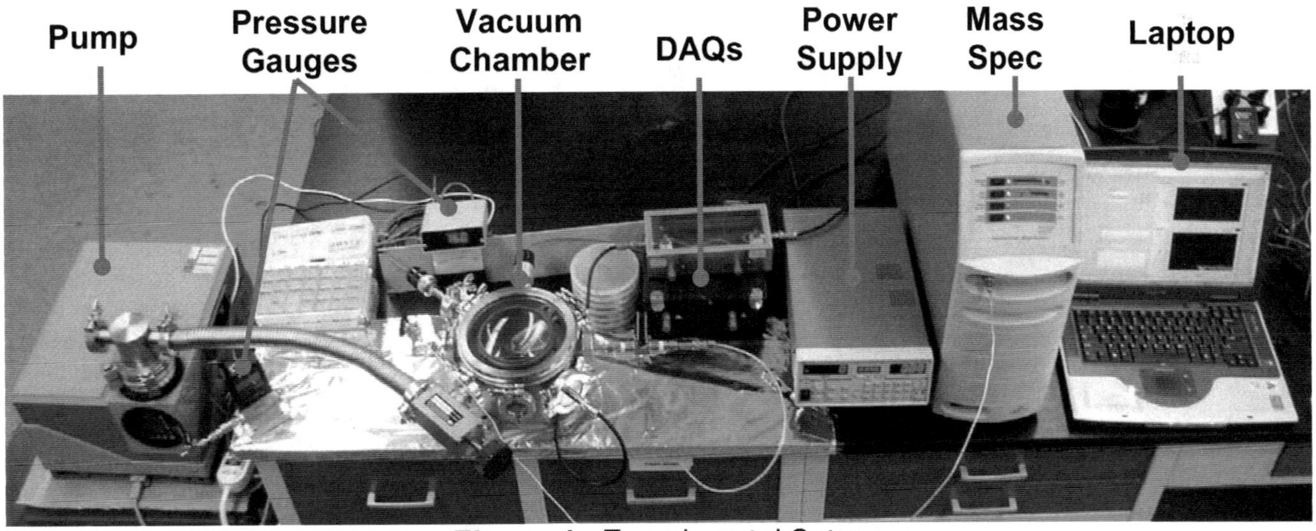

Figure 4. Experimental Setup

Developing butane plasma began our work on volatile hydrocarbons. Butane experiments gave promise to developing a plasma with hydrocarbon fuels. Experiments with butane determine the effects of microplasma device geometries.

Results

Butane has been tested with various geometries to determine the effect of the device. Tests were done by limiting the applied current while monitoring the current and voltage. These tests sustained point plasmas of butane that exceeding power densities of 10MW/cm^3. Figure 5 shows an example of data from a butane plasma experiment.

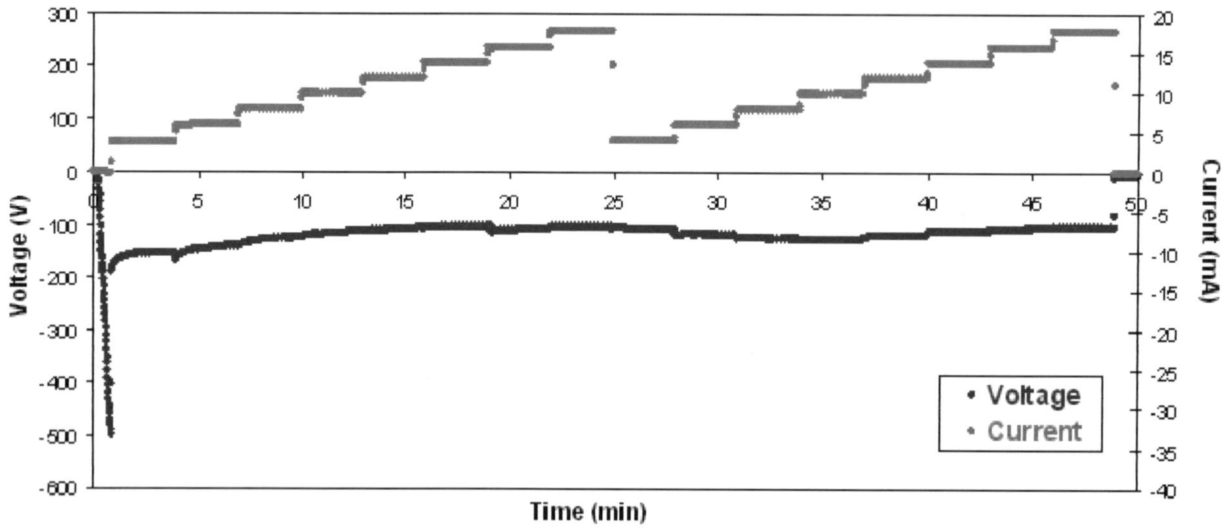

34

Figure 5. Microplasma Butane at 760 torr

These experiments give us the ideal dimensions for the second generation micro fluidic reactors.

Additional tests to seal the current plasma devices have shown varying effects that must be developed to be reliable prior to the next evolution of the microplasma reactor. Second-generation enclosed microplasma reactors will allow for a continuous feed-flow reaction while concealing the plasma. These reactors will first be experimented with butane under varying reaction types. After which key components of JP-8 will then be tested, requiring the addition of a vaporizer. These experiments will move us closer to the final goal of reforming JP-8 to provided portable power on the battlefield. The results of the experiments discussed above will be presented at the conference in detail.

Acknowledgements

The authors wish to express their gratitude to the Army Research Office for supporting this project and to various individuals which contributed to the experiments (E. Whittaker, V. Tarnovsky, G. Wohlrab, B. Fraser, A. Najem, K. Ciampi, E. Lennon, J. Hwang, A. Omosebi, J. Weinstein, T. Heithoff, S. Hwang, J. Symonds).

References

1. G. Petitpas, J.-D. Rollier, A. Darmon, J. Gonzalez-Aguilar, R. Metkemeijer, L. Fulcheri, "A comparative study of non-thermal plasma assisted reforming technologies," *International Journal of Hydrogen Energy*, **32**, 2848-2867 (2007).
2. T. Paulmier, L. Fulcheri, "Use of non-thermal plasma for hydrocarbon reforming," *Chem. Eng. J.,* **106**, 59-71, (2005).
3. M.V. Mundschau, C. G. Burk, D. A. Gribble Jr., "Diesel fuel reforming using catalytic membrane reactors," *Catalysis Today*, **136**, 190–205, (2008).
4. Lt. Col. John LaTulip, "Lessons Learned from Operation Iraqi Freedom," *Tri-Services Power Expo*, Norfolk, VA, (2003).
5. D. R. Cohn, L. Bromberg, K. Hadidi, "Onboard Plasmatron Hydrogen Production For Improved Vehicles," *MIT Plasma Science and Fusion Center*, (2006).
6. Wikipedia contributors, "Plasma (physics)," *Wikipedia, The Free Encyclopedia*, (2008). Available at: http://en.wikipedia.org/w/index.php?title=Plasma_(physics)&oldid=233794520.
7. Wikipedia contributors, "Fusion Energy," *Wikipedia, The Free Encyclopedia,* (2008). Available at: http://en.wikipedia.org/w/index.php?title=Fusion_power&oldid=233899672.
8. H. Scott Fogler, "Elements of Chemical Reaction Engineering 3rd Edition", *Prentice Hall PTR*, (1999).
9. Wikipedia contributors, "Activation Energy," *WIkipedia, The Free Encyclopedia*, (2008). Available at: http://en.wikipedia.org/w/index.php?title=Activation_energy&oldid=231344369.
10. B. Cannela "Refining and Distillate Fuel Trends, " 4th *DoD Logistics Fuel Reforming Conference*, October 22-23, Philadelphia, PA, (2003).
11. D. J. Moon, J. W. Ryu, K. S. Yoo, D. J. Sung, S. D. Lee, "Development of iso-octane fuel processor system for fuel cell applications," Catalyst Today, **136**, 222–227, (2008).
12. K. Becker, Final Report, "Compact Plasma Reactor for Ship Service Fuel Cell System," *Office of Naval Research Ship Service Fuel Cell Program*, pg 234, (2004).
13. D. Dietz, H. Ghezel-Ayagh, J. Hunt, A. Belkind, K. Becker, and A. Nickens, "Plasma Treatment of a Heated Diesel/Steam Mixture for Use in Ship Services Fuel Cell Systems", *Proc. 31st IEEE ICOPS*, Balitmore, MD, pg.412, (2004).
14. R. Foest, M. Schmidt, K. Becker, "Microplasmas, an emerging field of low-temperature plasma science and technology," International Journal of Mass Spectrometry, **248**, 87-102, (2006).
15. X. Ouyang, L. Bednarova, P. Ho, and R.S. Besser, "Preferential Oxidation of Carbon Monoxide in a Thin-Film Catalytic Microreactor: Advantages and Limitations," *AIChE J.,* **51** (6), 1758-1772, (2005).
16. J. Reece Roth, "Industrial Plasma Engineering", *Institute of Physics Publishing, Bristol and Philadelphia*, (1995).

219c Miniature Reformer Design for Autothermal Reforming of Methanol to Generate Hydrogen for Portable Power Applications

Vinay K. Vadlamudi and **Srinivas Palanki**. Chemical Engineering, University of South Alabama, 307 Univ. Blvd N, EGLB 244, Mobile, AL 36688

In this research a miniaturized packed bed reactor is designed and analyzed, in which autothermal reforming of methanol occurs to produce sufficient hydrogen for generating 50 W of power. Mass and energy balance equations are developed for each species in the reactor. The pressure drop is modeled via the Ergun equation. Simulations are conducted in MATLAB to determine the effect of process parameters (e.g. steam to methanol ratio, oxygen to methanol ratio, inlet pressure, inlet temperature) on the production of hydrogen. For a range of catalyst particle sizes considered, it is shown that the pressure drop is negligible. An appropriate reactor jacket is designed where methanol is oxidized to produce sufficient heat for the steam reforming reactions in the reactor. The flow rate of methanol in the jacket is computed so that sufficient heat is generated for complete conversion of methanol under non-isothermal conditions. Calculations are done for the energy required to bring the fuel processor system from ambient temperature to the operating temperature where fuel cell quality hydrogen can be produced. Effectiveness factor calculations are performed on a range of catalyst particle sizes. Finally, the simulation results are compared with experimental results from the literature.

219d Desulfurization of Liquid Fuels by Adsorption for Fuel Cell Applications

Ambalavanan Jayaraman, Gokhan Alptekin, Margarita Dubovik, Matthew Schaefer, John Monroe, and Kristin Bradley. TDA Research Inc., 12345 W. 52nd Ave, Wheat Ridge, CO 80033

Power generation by portable fuel cells require a source of hydrogen. Liquid fuels are a suitable source due to their availability and high hydrogen content. The major drawback to the use of liquid fuels in portable fuel cells as a source of hydrogen is their sulfur content. EPA regulations have lowered the sulfur level in the commercial fuels to 30 ppmw S for gasoline and 15 ppmw S for diesel, this makes them ideal fuel for fuel cells in transportation and portable applications but the sulfur level in the logistic fuels is still high 3,000 ppmw S for jet fuels (JP-8, JP-5) and 10,000 ppmw S for naval distillate (NATO F-76). The effective utilization of transportation fuels in fuel cell applications requires removal of refractory sulfur species (organosulfur compounds) to below 0.1 ppm. Low temperature fuel cells (e.g. PEM) require clean (essentially pure) hydrogen feed to prevent the poisoning of the anode catalyst. Even the more robust high temperature fuel cells (e.g., solid oxide fuel cells) are poisoned with low levels of sulfur contaminants. Sulfur removal is critical for fuel cells and adsorption is a promising technology for accomplishing such low levels of sulfur. TDA has developed a regenerable sorbent-based fuel desulfurization system that can easily integrate with any fuel cell fuel processor. TDA's adsorbent removes all of the refractory organic sulfur compounds from the fuel, which is still in the liquid phase and reduces the total fuel sulfur content to sub-ppm levels (e.g., less than 0.1 ppmw). For portable fuel cell applications TDA's adsorbent can be used in an expendable mode or in a centralized multi-bed regenerable desulfurization system. TDA has built a four-bed prototype liquid fuel desulfurization system, which will be tested with various liquid fuels. The test results from the prototype unit testing and the cyclic regeneration and stability data for TDA's SulfaTrap™ –D1 (TDA's liquid fuel desulfurization sorbent) will be discussed in this paper.

Improved Thermal Management of H_2O_2 Decomposition with Microreactors for a SOFC Oxidant Source aboard Unmanned Undersea Vehicles

E. Lennon[1], A. A. Burke[2], and R. S. Besser[1]
1.Stevens Institute of Technology, Hoboken, NJ 07030 (elennon@stevens.edu, rbesser@stevens.edu)
2. Naval Undersea Warfare Center, Newport, RI 02841

Abstract

The catalytic decomposition of hydrogen peroxide (H_2O_2) into water (H_2O) and oxygen (O_2) [$H_2O_2 \rightarrow H_2O + \frac{1}{2} O_2$] is notoriously susceptible to thermal runaway (Heat of Reaction: -98 kJ/mol). However, liquid H_2O_2 has high oxygen density making it a particularly attractive oxidant for applications where air-independence and space limitations elicit additional power supply design challenges. Microchemical technologies have intrinsically higher surface-to-volume ratios (SV) potentially facilitating heat and mass transfer. This study investigated a micro-scale reactor to illustrate thermal management and assess oxygen production during the multiphase decomposition of H_2O_2 as an oxidant for an unmanned undersea vehicle (UUV) solid oxide fuel cell (SOFC). A microscale packed bed (MPB) reactor model with channel radius 0.5 mm was developed in the finite element modeling program COMSOL. The properties of the multiphase product stream were described by weighted averages of the component species under the assumption of a well-mixed, dispersed phase. The micro-scale channel alone (SV: 2254 m^2/m^3) failed to prevent significant heat rise during simulated decomposition in the convectively cooled reactor channel ($T_{rise} \sim 100$ K). The extension of surface area around the reactor channel (SV: 188,439 m^2/m^3) improved passive cooling, resulting in simulated MPB temperature rises less than 7 K and experimental MPB temperature rises less 4 K. Decomposition was also tested using a second miniature tubular, catalyst-coated (TCC) reactor model with channel radius of 1.5 mm (SV: 2666 m^2/m^3) for thermal comparison with the MPB. Experimental data collected with the MPB reactor were compared to the numerical MPB models. Pulsating oxygen production, consistent with two-phase slug flow, was observed at the outlet of the MPB reactor over flow rates of 0.05 and 0.15 ml/min. Although slug flow was not specifically accounted for in the model, the time-averaged empirical behavior compared well to the simulated behavior for the MPB. Thermal management of the MPB was found to be superb and clearly superior to that of the miniature scale TCC. Overall these results showed the feasibility of oxygen generation and improved thermal control during H_2O_2 decomposition with a microchemical approach relative to conventionally-sized or even miniature (mm-scale) reactors.

Introduction

A major challenge facing the design of air-independent power systems is continuous energy generation. High energy density and potentially low operating costs make fuel cells an attractive option for high endurance, undersea vehicle applications [1]. However, undersea fuel cell power systems require an oxidant supply. Hydrogen peroxide (H_2O_2) readily decomposes into water (H_2O) and oxygen (O_2) [$H_2O_2 \rightarrow H_2O + \frac{1}{2} O_2$] upon contact with a number of commercially available metal catalysts offering a dense source of oxygen per unit volume that makes it a valuable commodity as an air-independent fuel cell oxidant [2]. Kinetic experiments have shown manganese dioxide (MnO_2) is an effective and inexpensive H_2O_2 decomposition catalyst with reported activation energies ranging from 20 to 44 kJ/mole [3]. Furthermore, the high oxygen density, ease of handling, and commercial infrastructure of H_2O_2 add to its appeal as an oxidant source [4].

Despite these advantages, it is well established that catalytic H_2O_2 decomposition is highly exothermic, which has historically limited its application as a power system oxidant, especially within the US Navy community [5]. In addition, like other exothermic multiphase reactions, the liquid and gas product stream from H_2O_2 decomposition can lead to undesirable pressure increases and an irregular distribution of thermal capacity in conventionally sized reactors promoting hot spots and the possibility of thermal runaway [6]. Microscale (relevant fluid dimension in subunits < 1 mm) chemical systems by virtue of small-scale geometry, possess high surface-to-volume ratios resulting in heat and mass transfer coefficients capable of inhibiting thermal runaway. Conventional reactors have surface to volume ratios of 10^2 m^2/m^3, whereas small-scale reactors possess surface-to-volume ratios that are at least an order of magnitude greater [7]. The microscale packed bed (MPB) microreactor and tubular catalyst coated (TCC) reactor used in this research had surface-to-volume ratios of $1.8x10^5$ m^2/m^3 and $2.6x10^3$ m^2/m^3 respectively. Although laminar flow is typical in many homogenously dispersed product streams within microchannels [8], slug flow may also develop. One study presented two phase flow evolution of O_2 gas bubbles and liquid water from platinum catalyzed H_2O_2 decomposition in 80 μm diameter microchannels in three stages: liquid decomposition, bubble coalescence, and bubble slug [9]. Despite the complexities of the multiphase product stream generated from H_2O_2 decomposition, this work aimed to demonstrate thermal management feasibility and assess O_2 production during the catalytic multiphase decomposition of H_2O_2 of the MPB reactor.

Methods

Simulated H_2O_2 Decomposition

I. Model Geometries

Figure 1A shows the MPB channel with a height of 0.5 mm and length of the 5 cm. Surrounding the MPB channel was a rectangular stainless steel 316 block (7x2x0.4 cm). A Plexiglas (PMMA) cover (7x2x0.1 cm) served as the reactor seal. Shown in Figure 1B is the TCC geometry and dimensions in cm used for thermal comparison with the MPB.

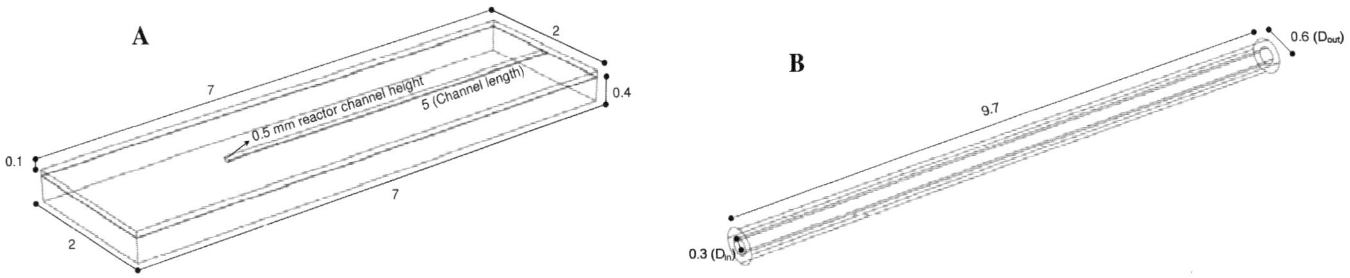

Figure 1A and 1B: MPB Model Geometry and TCC model geometry. All dimensions in cm except where noted.

II. Balance Equations and Assumptions

Mass Balance. Assumptions pertaining to the phases of the reactant and product species of the mass balance were made. The amount of H_2O_2 in the vapor phase was considered negligible. This assumption was supported by the fact that equilibrium vapor concentration of 50% w/w H_2O_2 reactant at 333 K is less than 10% [3]. Additionally, the amount of dissolved O_2 in the liquid was considered negligible compared to the H_2O_2 and water liquid components, and the amount of water vapor in the gas phase was considered negligible compared to the O_2 gas provided the temperature rise remained below the boiling temperature of water (373 K). The mass balance included the diffusion of H_2O_2 into liquid water and H_2O_2., the bulk flow of the evolving fluid mixture (consisting primarily of liquid H_2O_2 solution in water, and O_2 gas) down the reactor channel, and consumption of H_2O_2 resulting from the decomposition reaction. To model the mass balance, the built-in convection and diffusion application mode in the program COMSOL was used. Equation 1 governed the convection and diffusion mass balance of H_2O_2.

$$\nabla \cdot \left(-D_{H_2O_2eff} \nabla C_{H_2O_2}\right) + u_s \cdot \nabla C_{H_2O_2} = r_{H_2O_2} \quad \textbf{(1)}$$

In equation 1, $D_{H2O2eff}$, was the effective diffusivity of H_2O_2 into the liquid water, H_2O_2 solution. The diffusivity term neglected diffusion interactions with the catalyst. The mass balance solved for the concentration, C_{H2O2}. The variable u_s represented the superficial velocity of the fluid. The rate of reaction based on the consumption of hydrogen peroxide, r_{H2O2}, was established assuming first order kinetics weighted via an inputted catalyst mass. The MPB activation energy was set to 20 kJ/mol, whereas the TCC activation was set to 40 kJ/mol. The water and oxygen production rates were related to r_{H2O2}. A conventional Arrhenius relationship defined the reaction rate constant coupling it to the energy balance (equation 2).

Energy Balance. The energy balance of the microreactor system included conduction, advective input and output due to the feed and exit streams, exothermic heat generation due to reaction, sensible heat, and the heat of water vaporization. To model the energy balance, the built-in convection and conduction application mode in COMSOL was used. Equation 2 governed the convection and conduction energy balance.

$$\nabla \cdot \left(-k_{eff} \nabla T\right) + u_s c_p \rho \cdot \nabla T = \Delta H_{rxn} r_{H_2O_2} - S_H + \Delta H_{vap} r_{H_2O} \quad \textbf{(2)}$$

In equation 2, k_{eff} gave the effective thermal conductivity of the reactor channel. The energy balance solved for the variable temperature, T, coupling it to the mass balance equation. The term, u_s, still represented the superficial velocity of the fluid. The term c_p represented heat capacity and the term, ρ, the average density of the fluid in the microchannel. The heat of reaction, ΔH_{rxn}, defined the heat released during the exothermic decomposition reaction. The sensible heat term, S_H, described the contribution of the heat capacities per product species required to increase the temperature to the boiling point of water. The remaining term of the energy balance, ΔH_{vap}, expressed the energy required to vaporize water based on the rate of water production during the reaction.

Momentum Balance**.** To model the momentum balance, the built-in non-isothermal flow application mode in COMSOL was used. The non-isothermal flow application solves the Navier-Stokes equations for weakly compressible flows (flows with Mach numbers < 0.3). Neglecting the effects of the catalyst in the flow channel and using low initial flow rates maintained the conditions defining weakly compressible fluid flow. The density, ρ, was modeled as an average of the liquid and gas comprising the fluid weighted by the mole fractions of H_2O_2, H_2O, and O_2 species respectively and changed as the reaction progressed (equation 3).

$$\rho = y_{H_2O_2}\rho_{H_2O_2} + y_{H_2O}\rho_{H_2O} + y_{O_2}\rho_{O2} \quad \textbf{(3)}$$

In equation 3, y_i represents the total mole fraction of component i (H_2O_2, H_2O or O_2). This average density was applied under the simplifying assumption that the generated O_2 gas was homogenously dispersed throughout the effluent liquid. The momentum balance solved for the velocity field giving the superficial velocity of the fluid and the pressure.

Experimental H_2O_2 Decomposition

Effluent oxygen gas rates and temperature data from the MPB experimental microreactor were collected according to the schematic in Figure 2. A syringe or micro - pump was used to dispense thirty or fifty percent H_2O_2 weight in water (30% w/w and 50% w/w) reactant at volumetric flow rates (VFR) from 0.05 ml/min up to 0.15 ml/min. Analogous conditions were used to collect data from the TCC for thermal comparison. The experimental reactor was connected to a water trap system which facilitated the separation of the oxygen gas and liquid product. The O_2 gas proceeded to the oxygen flow meter and the liquid product flowed to the glass collection vial. A NI USB 6008 digital-to-analog converter transmitted the flow data to the laptop for collection and storage. Four thermocouples were placed in 10 mm increments down the length of the MPB reactor channel. A National Instruments USB 9211 digital to analog converter connected to 9162 USB data acquisition device designed specifically to interface with thermocouples transmitted the temperature data to the laptop for temperature data collection and storage. The program Labview Signal Express controlled both the temperature and flow data recording processes.

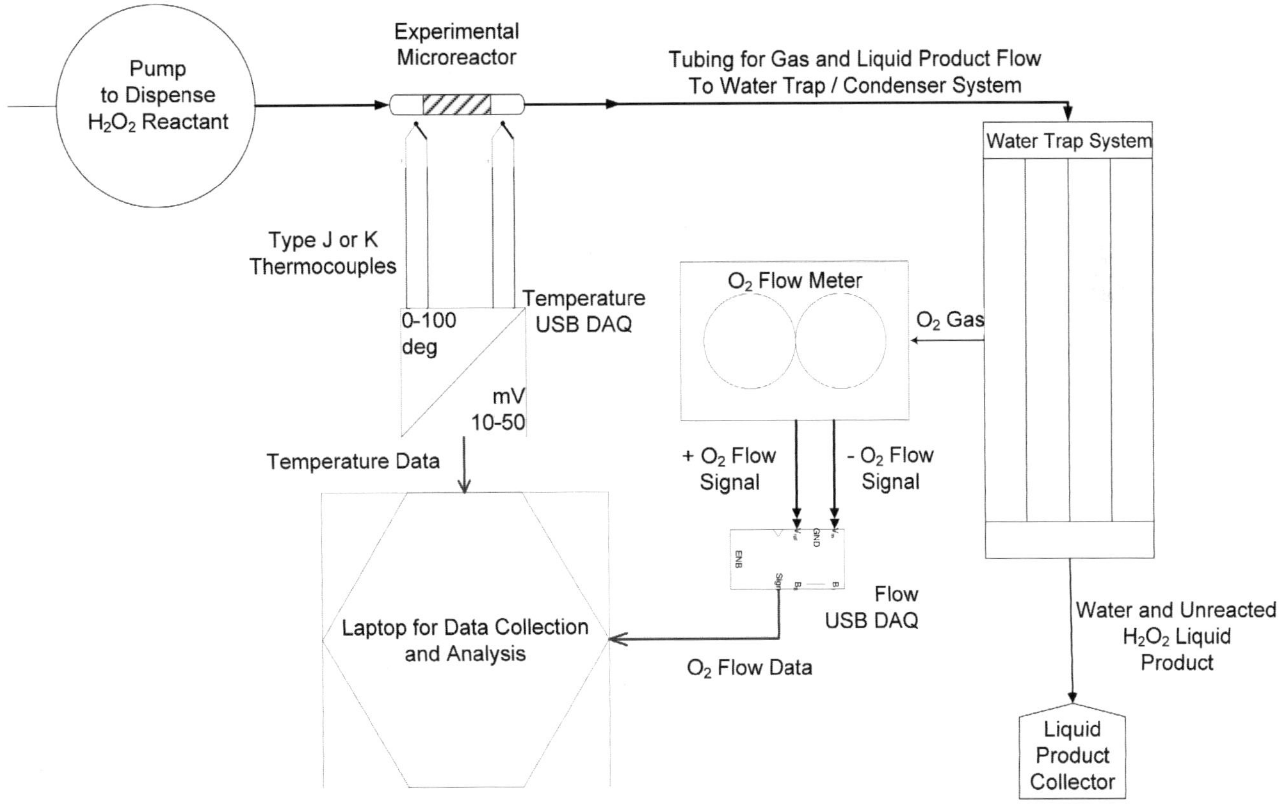

Figure 2: Experimental Setup Schematic

Results and Discussion

The three-dimensional steady state temperature profile and corresponding H_2O_2 concentration profile under the stated conditions for the extended surface area MPB model is shown in Figure 3. Simulations were run to determine which flow rates and concentrations were necessary for 30% or greater conversion given the MPB geometry. The maximum simulated steady state temperature rise for the MPB was less than 10 K for the 50% reactant case with a VFR of 0.15 ml/min, the case that would theoretically generate the greatest heat for the given MPB geometry. Extending the surface areas around the reactor channel further increased the SV ratio of the MPB reactor resulting in enhanced convective cooling, while yielding a simulated steady state conversion greater than 50%. This prompted use of experimental flow rates between 0.05 ml/min and 0.15 ml/min for 50% w/w H_2O_2 reactant concentrations. The experimental MPB reactor during a representative decomposition experiment is shown in Figure 4. In the image of the experimental MPB, there is clearly multiphase product in the channel.

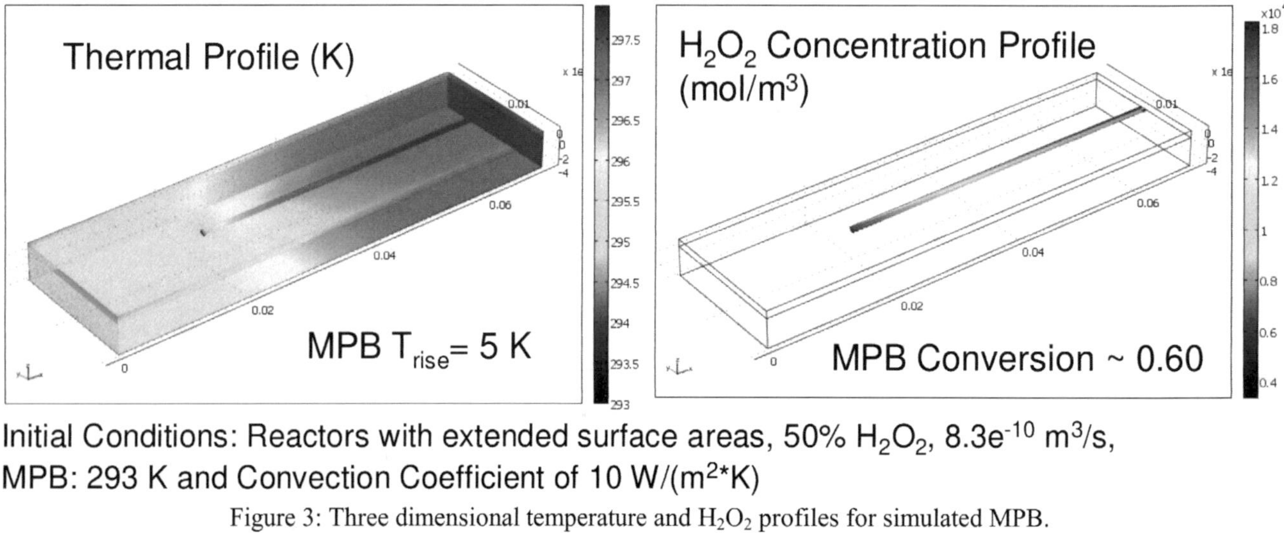

Figure 3: Three dimensional temperature and H_2O_2 profiles for simulated MPB.

Figure 4: Experimental MPB showing multiphase product stream.

To compare the thermal management of experimental MPB with the simulated MPB, 4 thermocouples recorded temperature at 1, 2, 3, and 4 cm down the length of the experimental MPB. Figures 5A and 5B show the simulated steady state MPB temperature profile versus reactor length given by the solid trend line and the average temperature recordings for 3 experimental runs under steady state operation of the MPB using 50% w/w H_2O_2 at an initial VFR of 0.05 ml/min and 0.15 ml/min respectively. To compare the thermal management of the MPB with the TCC, 2 thermocouples recorded temperature at 3 and 6.7 cm down the length of the 9.7 cm long TCC reactor. Figures 6A and 6B show the simulated steady state TCC temperature profile versus reactor length given by the solid trend line and the average temperature recordings for representative experimental runs under steady state operation of the TCC using 50% w/w H_2O_2 at an initial VFR of 0.05 ml/min and 0.15 ml/min respectively.

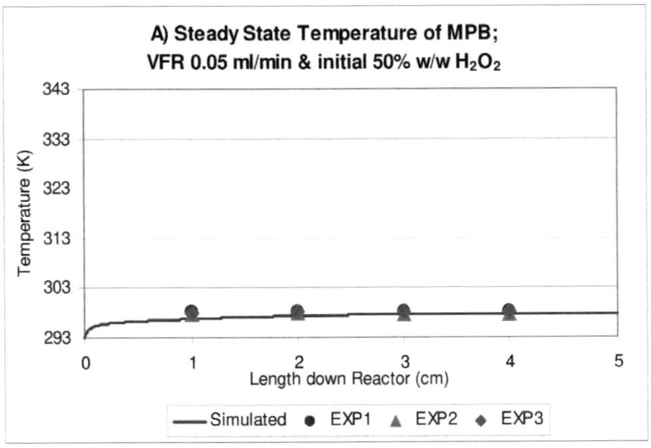

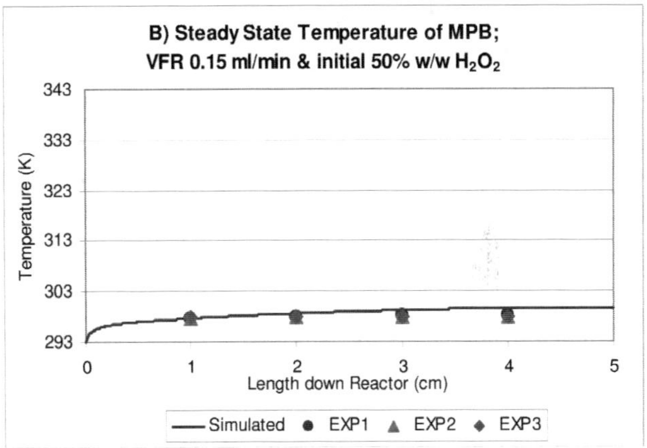

Figures 5A and 5B: Simulated and recorded steady state MPB temperatures down reactor length for a VFR of A) 0.05 ml/min and B) 0.15 ml/min.

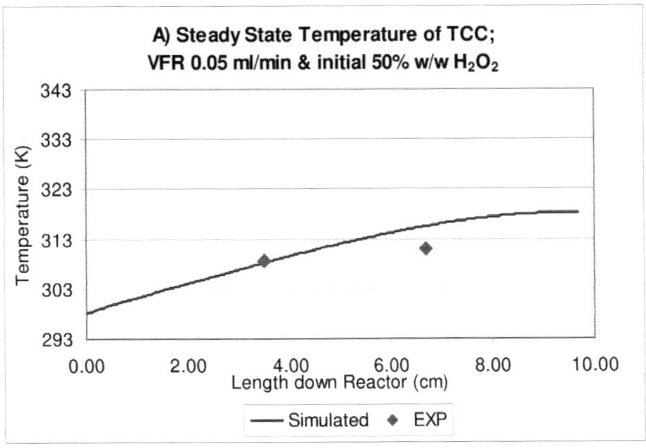

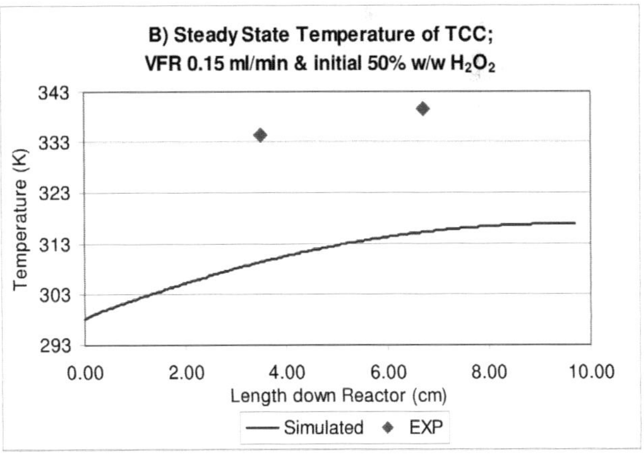

Figure 6A and 6B: Simulated and recorded steady state TCC temperatures down reactor length for a VFR of A) 0.05 ml/min and B) 0.15 ml/min.

Figures 5A and 5B show that the temperature rise was less than 10 K for both the simulated and experimental MPB. Figures 5A and 5B also show that the experimental MPB readings were close to those simulated. However, the experimental MPB data at the higher flow rate had slightly lower temperatures than those simulated. This is likely due to lower experimental conversion compared to the conversion predicted during simulation under the same conditions (Figure 7). When comparing the MPB to the miniature TCC temperatures in Figures 6A and 6B, the MPB shows superior thermal control. Although the TCC exhibited temperature rises less than 100 K, which may be possible with conventionally sized high concentration H_2O_2 decomposition reactors, it still reached a maximum steady state, experimental temperature of 47 K near the TCC outlet, and was greater than the steady state temperature trend predicted during simulation. This deviation, which was larger at the higher TCC VFR, relates to the break down of the homogenously dispersed, well-mixed assumption used in the modeling and also contributed to the deviation between the simulated and experimental TCC outlet conversions. Although the break down of the well-mixed assumption at the higher VFR affects both the MPB and TCC reactors, the MPB maintained thermal control over the reaction zone, indicating the high MPB SV ratio was successful in mitigating unwanted temperature rise.

Figure 7 depicts the steady state outlet conversions for the MPB and TCC at the stated VFRs using an initial 50% w/w H_2O_2 and corresponds to the temperature simulations and recordings in Figures 5 and 6 for the given conditions. The highest conversion was observed in the simulated MPB at the lower VFR. This was expected because the lower VFR in the microchannel resulted in the greatest residence time. However, the experimental MPB showed somewhat lower steady state conversion than the simulated values at both flow rates. This decrease in conversion can be explained by the intermittent occupation of catalyst sites by produced oxygen bubbles. These interactions with the catalyst were neglected in the model, but contribute to lower conversions during experimental runs. The TCC run at the higher VFR displays an exception to the mass transfer limitation of oxygen leaving the catalyst sites. In the experimental TCC, distinct slugs of gas and liquid were observed at the outlet, indicating the break down of the well-mixed assumption. The evolution of slug flow in the TCC reactor at the higher VFR probably caused the heat capacity to vary down the reactor channel, permitting the formation of localized hot spots, an increase in reaction, and subsequently higher overall conversion. This result emphasizes the superior control the MPB microreactor has over the prevention of hot spots, and thus the prevention of thermal runaway.

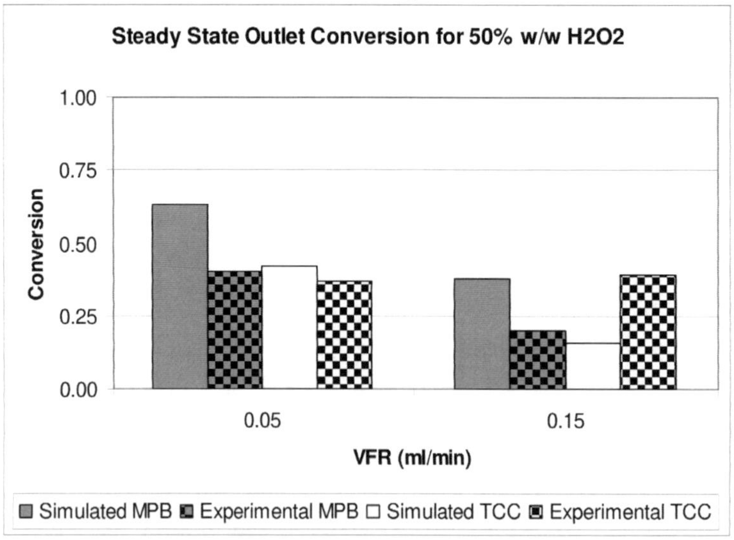

Figure 7: Steady state outlet conversions

Conclusions

Significant temperature rise (up to 100 K) occurred across a simulated microchannel reactor when the extended surface area was omitted. Increasing the surface-to-volume ratio to enhance passive convective cooling of the simulated MPB facilitated thermal management during catalyzed H_2O_2 decomposition. Thermal management was successfully accomplished with the addition of the high area heat exchanger in the simulated MPB reactor, subsequently mitigating the risk of thermal runaway. The momentum balances within the simulated reactors were included under the assumption of homogenously dispersed oxygen gas in a weakly compressible two-phase fluid to characterize the multiphase effluent product stream. Under this assumption, both the MPB and TCC models demonstrated the promise of thermally controlled H_2O_2 decomposition. However when the simulated and experimental data were compared, differences between the simulated and experimental TCC data were large at the higher VFR, indicating the assumption of homogenously dispersed flow broke down at the higher VFR, especially for the TCC, and the recorded temperature rise was greater than that simulated. Although the divergence of the TCC indicated the break down of the well-mixed assumption at the higher flow rate, and the possible production of hot spots, the consistently small temperature rise (<10K) of the MPB showed the mitigation of thermal runaway even at the higher VFR. However, the observed enhanced thermal control of the MPB may partially result from the temporary occupation of catalyst sites by oxygen gas bubbles, which serves to depress conversion slightly. This possibility is supported by the lower experimental conversions relative to simulated conversion in all cases, except for the TCC at the high VFR, which displayed higher susceptibility to heat effects. Therefore, ways of increasing the conversion via improved catalyst application techniques and the modification of flow dynamics which could dislodge bubbles should be pursued. These findings present a firm foundation for further optimization of thermally controlled H_2O_2 decomposition microscale reactors capable of achieving higher conversions, and increased oxygen generation.

References

1 The Navy Unmanned Undersea Vehicle (UUV) Master Plan. In Navy, U.S.D.o.t., ed, pp. 16, 55-58, 67, 702004).

2 Salem, I.A.E.-M., Mohamed; Zaki, Ahmed B. Kinetics and Mechanisms of Decomposition Reaction of Hydrogen Peroxide in Presence of Metal Complexes. *International Journal of Chemical Kinetics*, 2000, 32(11), 643-666.

3 Bramanti, C.T., L.; Cervone, A.; d'Agostino, L.; Musker, A. J.;, and Saccoccia, G. Experimental Characterization of Advanced Materials for the Catalytic Decomposition of Hydrogen Peroxide. *Proceedings of 42nd AIAA/ASME/ SAE/ASEE Joint Propulsion Conference & Exhibit* Sacramento, CA, USA, 2006).

4 Davis, N.S., and Keefe, J.H. Concentrated hydrogen peroxide as a propellant. *Industrial Engineering and Chemistry*, 1956, 48(4), 745–748.

5 Papadaki, M., Marqués-Domingo, E., Gao, J. and Mahmud, T. Catalytic decomposition of hydrogen peroxide in the presence of alkylpyridines: Runaway scenarios studies. *Journal of Loss Prevention in the Process Industries*, 2005, 18(4-6), 384-391.

6 Guo, J., Jiang, Y. and Al-Dahhan, M.H. Modeling of trickle-bed reactors with exothermic reactions using cell network approach. *Chemical Engineering Science*, 2008, 63(3), 751-764.

7 Ehrfeld, W.V.H.H., L.,. *Microreactors: New Technology for Modern Chemistry*, pp. 5-8 (WILEY-VCH, Weinheim, 2000).

8 Jensen, K.F. Microreaction engineering -- is small better? *Chemical Engineering Science*, 2001, 56(2), 293-303.

9 Xu, J., Feng, Y. and Cen, J. Transient flow patterns and bubble slug lengths in parallel microchannels with oxygen gas bubbles produced by catalytic chemical reactions. *International Journal of Heat and Mass Transfer*, 2007, 50(5-6), 857-871.

219f A Hybrid-Silicon Miniature Hydrogen Fuel Cell for Portable Power Generation

Robert D. Morgan[1], Kevin Y. Lin[1], Hee Soo Kim[1], Bogdan Gurau[1], Mark Shannon[2], and Richard I. Masel[1]. (1) Chemical and Biomolecular Engineering, University of Illinois at Urbana-Champaign, 600 S. Mathews Ave, RAL 204, Urbana, IL 61801, (2) Mechanical Engineering, University of Illinois, 1206 W Green St, Urbana, IL 61801

The objective of this talk is to describe a 1mm^2 hybrid hydrogen / self-breathing cathode fuel cell that produces a higher power output than other previously reported miniature hydrogen fuel cells. As background, MEMS based fuel cells have been proposed to replace batteries in distributed sensor networks, on chip power, and implantable devices. As of yet, MEMS based fuel cells give performance no better than miniature batteries [1-9]. Prior to our work, the best MEMS based hydrogen/air fuel cell only produced 70 mW/cm2 compared to 149 mW/cm2 in our device. Yu et al[2] showed 195 mW/cm2 by running pure oxygen on the cathode, but note that powers are 4-5 times higher with oxygen than with air. Our device shows the highest power output of those shown previously.

---

1. Pichonat, T. and B. Gauthier-Manuel, Microsystem Technologies, 2007. 13(11): p. 1671-1678.

2. Yu, J., et al,. Journal of Power Sources, 2003. 124(1): p. 40-46.

3. Modroukas, D., et al., Journal of Micromechanics and Microengineering, 2005, 15(9) S193.

4. Meyers, J.P. and H.L. Maynard, Journal of Power Sources, 2002. 109(1): p. 76-88.

5. Yeom, J., et al., Sensors and Actuators B: Chemical, 2005. 107(2): p. 882-891.

6. Zhiyong, X., et al., Journal of Micromechanics and Microengineering, 2006(10): p. 2014.

7. Pichonat, T. and B. Gauthier-Manuel, Microsystem Technologies, 2006. 12(4): p. 330-334.

8. Chu, K.L., M.A. Shannon, and R.I. Masel, Journal of the Electrochemical Society, 2006. 153(8): p. A1562-A1567.

9. Chu, K.L., M.A. Shannon, and R.I. Masel, Journal of Micromechanics & Microengineering, 2007. 17 (9): p. S243-S249.

286a Design and Characterization of a Vapor Feed Methanol Fuel Cell with a Flowing Electrolyte

Fikile R. Brushett and Paul J.A. Kenis. Department of Chemical & Biomolecular Engineering, University of Illinois at Urbana-Champaign, 600 S. Mathews Ave., Urbana, IL 61801

Direct methanol fuel cells (DMFCs) are one of the most promising alternative power sources for portable applications because of methanol's low cost, easy storage and high energy density. Despite extensive research, the development of DMFC technology is hindered by fuel crossover which limits the operation of direct liquid fuel cells to dilute feeds [1]. Here we will focus on the design and characterization of an air-breathing vapor feed methanol fuel cell (VFMFC) with a flowing electrolyte stream. In contrast to liquid feed DMFCs, passive vapor feed DMFCs can simultaneously maintain a low methanol concentration, minimizing fuel crossover, but increase overall system energy density, utilizing a highly concentrated liquid methanol reservoir [2]. Under ambient conditions, the high vapor pressure of methanol causes substantial evaporation, enabling cell operation with no ancillary power losses [3]. We have systemically designed the cell to balance mass transport from the liquid fuel reservoir to anode catalytic surface with the removal of both carbon dioxide and unreacted methanol.

Additionally, as opposed to vapor feed DMFCs with conventional polymer electrolyte membranes, our VFMFC utilizes a flowing electrolyte stream located between the anode and cathode gas diffusion electrodes. A constantly refreshing electrolyte stream eliminates water management issues, limits fuel crossover and enables media flexibility, acidic or alkaline [4]. Key advantages of operating the VFMFC in alkaline media are superior reaction kinetics for both methanol oxidation and oxygen reduction. In addition to significantly improving cell performance, the facile kinetics enable the use of cheaper non-precious metal catalysts such as Ag and Pt_3Co cathode catalysts in lieu of Pt catalysts [5]. Moreover, any carbonates formed at the anode-electrolyte interface are removed by the flowing alkaline stream.

In this presentation, we will illustrate the design and characterization of a media flexible VFMFC. We will investigate cell performance as a function of methanol concentration, electrolyte type, flow rate and concentration. We will also present on the effects of various passive barrier layers on cell performance.

References

[1] Z. Guo and A. Faghiri, J. Power Sources, 167, 2007, 378; A. Heinzel, V.M. Barragán, J. Power Sources, 84, 1999, 70; J.G. Lui, T.S. Zhao, R. Chen, C.W. Wong, Electrochem. Commun., 7, 2005, 288.

[2] M. Hogarth, P. Christensen, A. Hamnett, A. Shukla, J. Power Sources, 69, 1997, 125; H. Fukunaga, T. Ishida, N. Teranishi, C. Arai, K. Yamada, Electrochim. Acta, 49, 2004, 2123.

[3] H. Kim, J. Power Sources, 162, 2006, 1232.

[4] R.S. Jayashree, M. Mitchell, D. Natarajan, L.J. Markoski, and P.J.A. Kenis, Langmuir, 2007, 6871.

[5] W.P. Zhou, R.S. Jayashree, F.R. Brushett and P.J.A. Kenis, J. Electrochemical Society, manuscript in preparation.

286b Electro-Oxidation of Formic Acid Over Pd-Pt-Au Trimetallic Catalysts for Portable Direct Formic Acid Fuel Cells

Won Suk Jung[1], Hideo Daimon[2], Jonghee Han[1], and **Su Ha**[3]. (1) Fuel Cell Research Center, Korea Institute of Science and Technology, 39-1 Hawolgok-dong, Seongbuk-gu, 136-791, Seoul, South Korea, (2) Development & Technology Division, Hitachi Maxell Ltd., 6-20-1 Kinunodai, Tsukubamirai, Ibaraki, Japan, (3) School of Chemical Engineering and Bioengineering, Washington State University, PO Box 642710, Pullman, WA 99164-2710

In this study, we synthesized novel trimetallic catalyst, PdPtAu, and characterized it electrochemical activity toward formic acid. The PdPtAu was prepared with NaBH4 as a reducer and calcinated at 95°C. Once it was synthesized, its crystalline structure and the extent of average particle size were investigated by X-ray diffraction and transmission electron microscopy. For this work, we investigated electrochemical activity of homemade PdPtAu toward formic acid, and compare it with those of commercial Pd, Pt and Au nanoparticles via cyclic voltammetry (CV) and chronoamperometry (CA) studies. To investigate the stability of our novel catalyst for formic acid oxidation and compare it with commercial noble nanoparticles, we measured their currents as a function of time over 2 hours in 5 M HCOOH + 0.5 M H2SO4, while a constant voltage was applied on the working electrodes. Additionally the electrodes were stripped after they were operated for 2 hours in 5 M HCOOH to understand the effects of surface poisoning on different catalysts. From this work, we conclude that our novel catalyst, PtPdAu, has a higher stability than the Pd and a higher electrochemical activity than Pt in the formic acid. Moreover, we observed that the PdPtAu is resistant to non COads species comparing to the Pd.

286c Micro-Electro-Mechanical Systems Based Design and Fabrication of Miniature Silicon Fuel Cells

Kevin Y. Lin[1], Robert D. Morgan[1], Likun Zhu[2], Hee Soo Kim[1], Bogdan Gurau[1], Mark A. Shannon[2], and Richard I. Masel[1]. (1) Chemical and Biomolecular Engineering, University of Illinois, 600 S. Mathews Ave, RAL 204, Urbana, IL 61802, (2) Mechanical Science and Engineering, University of Illinois, 1206 W Green, 122 Meb, Urbana, IL 61801

Portable devices have become much more feature packed and require much more power than before. However, they have been facing a persistent problem as rechargeable battery technology hasn't kept pace with the rapid advancements. Bridging that gap may now be possible through fuel cells, a clean energy technology that produces electricity from the electrochemical reactions between hydrogen and oxygen.

Recent interests [1-4] have been on MEMS-based microfabrication rather than traditional manufacturing processes due to ease of miniaturization and reduction of cost. Traditionally, fuel cells are built with separate manufactured pieces and then assembled at the end. The process is very costly, time consuming, and with high variability. With microfabrication, instead of assembling the separate parts, the fuels cells are made by addition and subtraction on a base structure. This is exactly the same way current microelectronic devices are made.

In this presentation, we will illustrate the design and fabrication of new MEMS-based fuel cells. Our approach is to work on single wafer MEMS fabrication along with a hybrid proton conducting membrane: silicon base structure filled with Nafion® ionomer solution.

[1] J.D. Morse, International Journal of Energy Research, 31, 2007, 576-602

[2] W.X. Ge, P. Cong, Nami Jishu yu Jingmi Gongcheng/Nanotechnology and Precision Engineering, 5, 2007, 182-189

[3] S. Motokawa, M. Mohamedi, T. Momma, S. Shoji, T. Osaka, Electrochemistry Communications, 6, 2004, 562-565

[4] Y. Yamazaki, Electrochimica Acta, 50, 2004, 663-666

286d Nanoporous Separator to Minimize Fuel Crossover In a Direct Methanol Laminar Flow Fuel Cell

Adam S. Hollinger, Mechanical Science and Engineering, University of Illinois at Urbana-Champaign, 1206 W. Green Street, Urbana, IL 61801, Larry J. Markoski, INI Power Systems, 175 Southport Drive, Suite 100, Morrisville, NC 27560, and Paul J.A. Kenis, Chemical and Biomolecular Engineering, University of Illinois at Urbana-Champaign, 600 S. Mathews Ave., Urbana, IL 61801.

As next-generation portable electronic equipment continues to shrink in size and increase in capabilities, high energy density direct methanol fuel cells (DMFCs) have become an attractive option to meet the additional power demands. While promising, DMFC development has been limited by issues related to the polymer electrolyte membrane which separates the anodic and cathodic compartments. Polymer electrolyte membranes must be kept hydrated to facilitate proton transport, i.e. water management, and permeation of methanol through the Nafion membrane, i.e. methanol crossover, results in mixed potentials at the cathode.

To eliminate the membrane constraints of a DMFC, we have previously designed a direct methanol laminar flow fuel cell (LFFC) that replaces a polymer electrolyte membrane with liquid-liquid contact between a fuel and electrolyte stream. This liquid-liquid contact limits methanol crossover to diffusion, but still allows a large area for potential methanol crossover. In order to inhibit methanol crossover in currently-designed LFFCs, the fuel concentration boundary layer must be pressed close to the wall via high flow rates or more commonly, wide channels. In an effort to maintain small electrode-to-electrode distances, for less volume and higher specific energy, we have placed a nanoporous polymer at the liquid-liquid interface of a methanol LFFC. This so-called separator greatly minimizes the total area of the fuel-electrolyte interface, and hence, the area in which unreacted methanol molecules can crossover to the cathode.

In this presentation, we will compare the performance a methanol LFFC with and without a separator. Electrochemical impedance spectroscopy (EIS) data will be used to characterize the separator's added resistance to charge transfer and the separator's ability to decrease methanol crossover while still maintaining small electrode-to-electrode distances.

286e Sequential Flow Membraneless Microfluidic Fuel Cells with Porous Electrodes

Kamil S. Salloum, Mechanical and Aerospace Engineering, Arizona State University, Tempe, AZ 85287 and Jonathan D. Posner, Mechanical and Aerospace Engineering / Chemical Engineering, Arizona State University, Tempe, AZ 85287.

A novel convective flow membraneless microfluidic fuel cell with porous electrodes is described. In this fuel cell design, the fuel flows through the anode and across a gap to the cathode. An oxidant is introduced into the gap between anode and cathode and advects downstream of the cathode. This fuel cell differs from previous membraneless designs in that the fuel and the oxidant flow in series, rather than in parallel, enabling independent control over the fuel and oxidant flow rate and the electrode areas. The cell uses formic acid as a fuel and potassium permanganate as the oxidant, both contained in a sulfuric acid electrolyte. The results show that increasing the electrolyte concentration reduces the cell Ohmic resistance, resulting in larger maximum currents and peak power densities. Increasing the flow rate delays the onset of mass transport and reduces Ohmic losses resulting in larger maximum currents and peak power densities. A comparison between the sequential flow and other flow patterns is discussed.

346a Fuel Cell Systems: An Introduction for the Chemical Engineer

Donald J. Chmielewski, Chemical and Biological Engineering, Illinois Institute of Technology, Suite 127, Perlstein Hall, 10 West 33rd Street, Chicago, IL 60616

This tutorial is intended for anyone with a background in Chemical Engineering and aims to provide a foundation for those interested in fuel cell research. Among the great challenges facing upcoming generations is the construction of a new globally focused and sustainable energy infrastructure. Clearly, many researchers believe that fuel cell systems will play a large role in this development, as evidenced by the plethora of articles published on the subject each year.

538a Water Problems In PEM Fuel Cells

Zhong-Sheng (Simon) Liu, Qianpu Wang, Datong Song, Cheng Huang, Michael Eikerling, Kourosh Malek, Zetao Xia, Javier Gazzarri, and Marc Secanell. NRC Institute for Fuel Cell Innovation, 4250 Wesbrook Mall, Vancouver, BC V6T 1W5, Canada

Water is produced inside a PEMFC's cathode catalyst layer (O2+4H++4e- =2H2O), a nano-scale porous composite thin-film, and the rate of water-production is proportional to electrical current density. A further increase of the electrical current density requires a better understanding of how and why water phase-change, liquid water morphology, water content inside electrolyte materials, proton conductivity and active surface area, and water transport mechanisms are related to operating conditions, catalyst layers properties, gas diffusion layer properties and flow-channel surface properties. This presentation talks about our efforts to model complex mass water transport phenomena and their interplay with the microstructures of catalyst layers, and it also discusses the multidisciplinary challenges and research opportunities.

538b Water Transport Characteristics of Gas Diffusion Layer in a PEM Fuel Cell

Ashok S. Damle, Techverse, Inc., 124 Goldenthal Court, Cary, NC 27519 and Vernon Cole, CFD Research Corporation, 215 Wynn Drive, Huntsville, AL 35805.

Effective management of water produced in a Proton Exchange Membrane (PEM) fuel cell is essential to avoid deterioration of fuel cell performance due to flooding of the catalyst and gas diffusion layers (GDL) or by drying of the membrane over the full operating temperature and load range. Ineffective water management leads to liquid-phase water blockage and mass-transport-limited performance or decreased protonic conductivity in the membrane and catalyst/GDL due to dehumidification of the ionomer. The characteristics of the GDL and catalyst layers, membrane and gas flow fields in bipolar plates affect water management and operation with GDL media being the most critical component influencing water transport in PEM fuel cells. This paper will present experimentally determined water transport characteristics through GDL media. Capillary pressure-water saturation correlations of various grades of GDL media were determined by "direct water displacement" technique with applied hydrostatic head. Dry as well as wet media gas permeability of the GDL media was also determined. In general, inhomogeneous nature of media pore structure was found to lead to water breakthrough at isolated locations on GDL surface predominantly influencing water/gas transport.

3-D Modelling of a Proton Exchange Membrane Fuel Cell with Anisotropic Material Properties

P.C. Sui[1], Sanjiv Kumar[2], Ned Djilali[1]
[1]Institute for Integrated Energy Systems, University of Victoria, Victoria, BC, Canada
[2] Ballard Power Systems, Burnaby, BC, Canada

Abstract

Fuel cell performance is dependent on a strongly coupled interaction taking place between convection, diffusion, electrochemical reactions, membrane water transport and heat transfer. Furthermore, the thermal and electrical properties of materials used in fuel cells are anisotropic. In addition, during normal operation of the fuel cell, the two phases of water are present both in the porous gas diffusion layers and in the channels of the oxidant and the fuel. To properly model the effect of all these phenomena in three dimensions with all the couplings and the anisotropy of the material properties presents a significant challenge. In this paper, aspects of a CFD-based modelling approach will be presented. Such an approach takes into account the coupling of the above phenomena by making some engineering assumptions in handling of the individual phenomenon. It is also able to treat the material anisotropy and different material properties used in the fuel cell and thus can be used as an engineering tool for understanding the impact of material changes in future designs. Typical results from these coupled 3-D simulations, which account for the anisotropic material properties, are presented. It is concluded that anisotropic properties play an important role in the coupled transport in a PEMFC and thus accurate measurement of these material properties for input in computational models is needed.

Introduction

Fuel cells are being pursued aggressively as a promising source of clean and renewable power as peaking of oil reserves looms in not so distant future. Several companies and organizations are developing fuel cell solutions for a range of applications of which stationary power generation and automotive applications are major components. Each application tends to have its own unique requirement for fuel cell design and operation. As the cost of development is a common thread in all of these, cutting design cycle times and getting the design right the first time continues to be a major emphasis for all developers. This has increasingly led to aggressive use of modeling, analysis and simulation tools within the industry. These tools at various levels of complexity are being used to study and improve concepts to 3-D designs and further aid in the analysis and troubleshooting of failure mechanisms as brought out by testing. Given that several multi-physical and multi-scale phenomena taking place in a fuel cell are still not fully understood, engineers have built heuristic models for these where needed while universities and other organizations continue to improve on the models to better account for the physics. The objective of this paper is to report on the 3-D modelling of a PEMFC using a simulation tool with anisotropic transport properties for the fuel cell components.

CFD-based Simulation Tools

Fuel cell modeling approach based on computational fluid dynamics (CFD) offers the engineer the advantage of studying and improving 3-D designs during the design cycle thus ensuring that enough effort is expended to minimize surprises down the line. Typically such CFD-based model solves for the governing conservation equations for mass, momentum, species and energy, which besides being a common backbone to many of the engineering problems, is also a relatively mature subject. Built on this CFD backbone are the additional models for electrochemical reactions, membrane water transport, electronic and ionic potential, and a saturation model to account for the two phase flow in PEMFC. These models are continuously being improved and are being calibrated using test data. The methodology of employing CFD for PEMFC modelling can be found in [2], among many others. The major advantage of using a comprehensive, 3-D CFD tool for PEMFC modelling over reduced dimension or simplified models is the high fidelity of the computational results. Although the computational time for CFD is often demanding, large scale parallel simulations using computer clusters have been demonstrated. In addition to being a tool for analysis, CFD-based tools can be modified to become a design tool, cf. [7].

Anisotropic Properties of Materials Used in a PEMFC

Most of the components of a PEMFC have anisotropic transport properties. The gas diffusion layer (GDL) in particular shows significant differences in the transport of electricity and heat in the in-plane and the through-plane directions. The catalyst layer is a composite material that is constructed by two networks (electrolyte and carbon black respectively), which shows some degree of anisotropy due to fabrication processes and also possibly due to preferred material orientation. The bipolar plate that is made of graphite powder and resin binder has some flaky structure that makes its transport properties anisotropic. Among all these materials, the anisotropy in the GDL is believed to be the most influential because of its fibrous structure as well as the fact that it is subject to discontinuous property change on one boundary, i.e. on the side that is in contact with the bipolar plate, cf. [3]. Furthermore, part of the GDL on the GDL-bipolar plate interface is under compression and the transport through such region is complicated, cf. [1] and [8]. Owing to the complexity of coupled transport phenomena, analysis on the effects due to anisotropy in transport properties of the materials in a PEMFC is best assisted by employing a simulation tool.

Results and Discussion

The transport property for a thermal or electrical conduction equation takes the form of a tensor:

$$\bar{k} \equiv \begin{vmatrix} k_{xx} & k_{xy} & k_{xz} \\ k_{yx} & k_{yy} & k_{yz} \\ k_{zx} & k_{zy} & k_{zz} \end{vmatrix} \qquad (1)$$

For a conventional plate-and-frame type of membrane electrode assembly (MEA) with coordinates X and Z as the in-plane direction and Y as the through-plane direction, we assume the transport is limited in both planes, i.e. only diagonal terms in the matrix remain. For a PEMFC problem, the anisotropy should appear in the conservation of species (diffusivity), momentum (permeability), energy (heat conductivity) and potential (electrical conductivity).

The baseline case used for the present study is a CFD model based on Ballard hardware. A commercial CFD software, CFD-ACE+, is used in the present study. The software with its capabilities for fuel cell simulations solves a complete set of conservation equations, which includes solution of conservation of charged species, water transport across the membrane, as well as transport of liquid water in porous media. The properties and model parameters used are obtained from validation with experimental data from [5]. The electrical conductivity used in the baseline case is isotropic and set to be 200 $(\Omega m)^{-1}$. A sensitivity analysis of numerical predictions for a Ballard hardware due to these properties shows that predicted performance is more sensitive to electrical conductivity than to thermal conductivity. Figure 2 shows the cell voltage predictions as a function of the anisotropy ratio, i.e. the ratio of the in-plane value versus the through-plane value. The through-plane value is kept constant for these calculations. One can see as the in-plane electrical conductivity is increased, the cell voltage increases because the ohmic loss in the GDL is reduced. For the thermal conductivity, the in-plane values have negligible impact on the overall cell voltage until it is reduced significantly, at which conditions the cell temperature rises because the transfer of the heat generated in the MEA is limited by the through-plane conduction and results into a higher temperature of the cell. The cell voltage appears to increase with increased cell temperature primarily due to a high electrical conductivity of the membrane at higher temperature.

Figure 3 shows a cross section plot of electrical potential and current flow in the cathode side and the MEA. The current from the catalyst layer sees a narrower flow path as it approaches the land area where the bipolar plate is in contact with the GDL. The in-plane resistance for current is relatively higher than the through-plane resistance, therefore a high current is seen near the area underneath the land. When the in-plane conductivity is set to two orders of magnitude higher than the baseline value while keeping the same through-plane conductivity as in the baseline, the current distribution changes significantly, cf. bottom of Fig. 3. In this case because the in-plane conduction is no longer important, the conduction becomes like one-dimensional dominated by the through-plane conduction. Figure 4 shows the through-plane current profile on the GDL/catalyst layer interface for different degree of anisotropy of electrical conductivity. The current profile for the isotropic case appears to be distorted towards the land area. As the in-plane conductivity is increased, the current profile becomes less distorted. As the disparity between the in-plane and through-plane conductivity increases, the transport in the in-plane becomes less dominating. As a result, the distribution of current density in the GDL is more uniform. In fact, when the gradient of electrical potential in the lateral direction diminishes, the distribution of current density on the GDL/catalyst layer interface will be likely dictated by mass transport in the GDL, cf. [6].

Figure 5 shows the temperature field of the MEA for isotropic (top) and anisotropic thermal conductivities. For the isotropic case the thermal conductivity is 2 W/m-k. For the anisotropic case, the in-plane conductivity is set to 100 W/m-k while the through-plane value is kept as 2 W/m-k. One can see that for the anisotropic case the temperature gradient as well as the MEA temperature are both lower than the isotropic case. Figure 6 shows a comparison of predicted polarization curves using isotropic and anisotropic electrical conductivity. The anisotropic case has the same through-plane conductivity as the isotropic case while its in-plane conductivity is one order of magnitude higher than the isotropic case. The reduced ohmic loss in the in-plane direction of the GDL results in a slightly higher cell voltage.

Conclusions

In the present study we report on comprehensive, 3-D modelling using a CFD tool for PEMFC, which includes specification of anisotropic transport properties for the MEA. It is demonstrated in the numerical predictions that the cell performance is sensitive to the anisotropic electrical conductivity. For the cases tested for Ballard hardware, it is shown that when the in-plane electrical conductivity is increased, the through-plane current density predictions on the GDL/catalyst layer becomes less distorted by the presence of the land area and the cell voltage increases because of reduced ohmic loss. It is also shown that the numerical prediction for cell voltage is not sensitive to the anisotropic thermal conductivity until the in-plane conductivity is substantially reduced. The present study indicates the importance of anisotropic transport properties and calls for measurement data for the materials used in the PEMFC.

Acknowledgement

PCS would like to acknowledge the funding provided by the MITACS Network of Centres of Excellence and Ballard Power Systems.

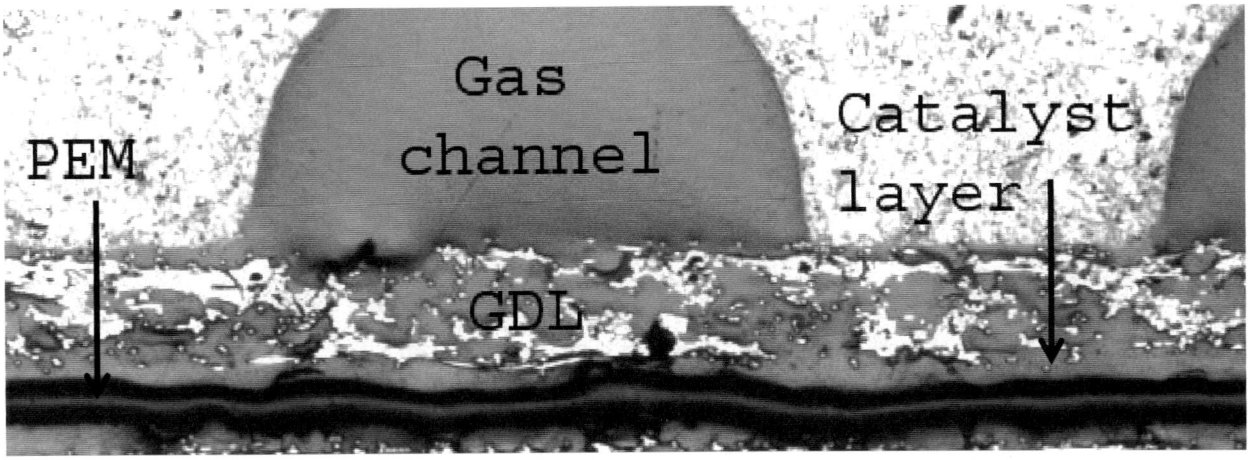

Figure 1. Cross sectional cut of an MEA

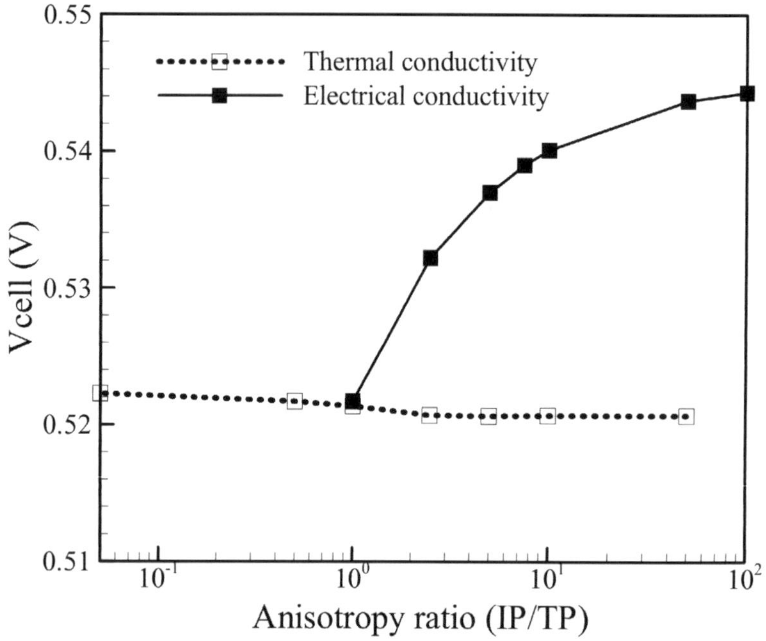

Figure 2. Sensitivity of predicted cell voltage due to anisotropy of electrical and thermal conductivities (IP= in-plane, TP= through-plane)

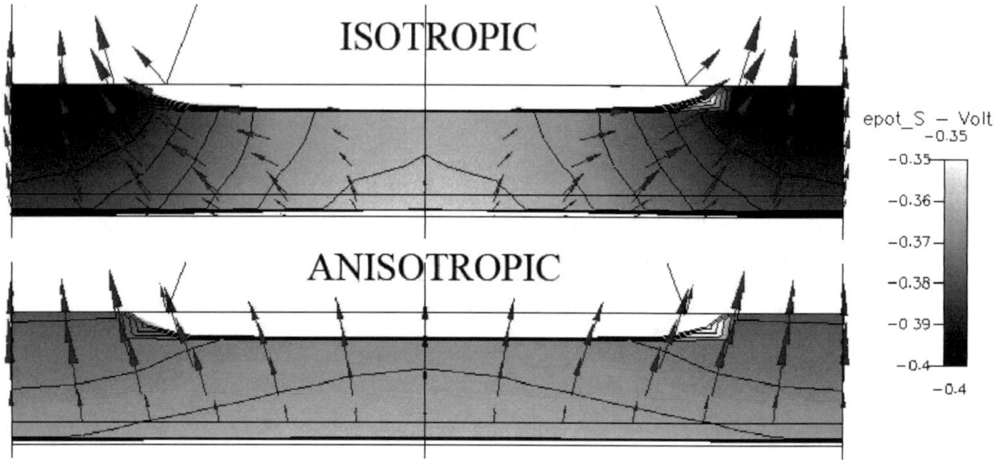

Figure 3. Electrical potential distribution and current vectors in the cathodic catalyst layer and GDL

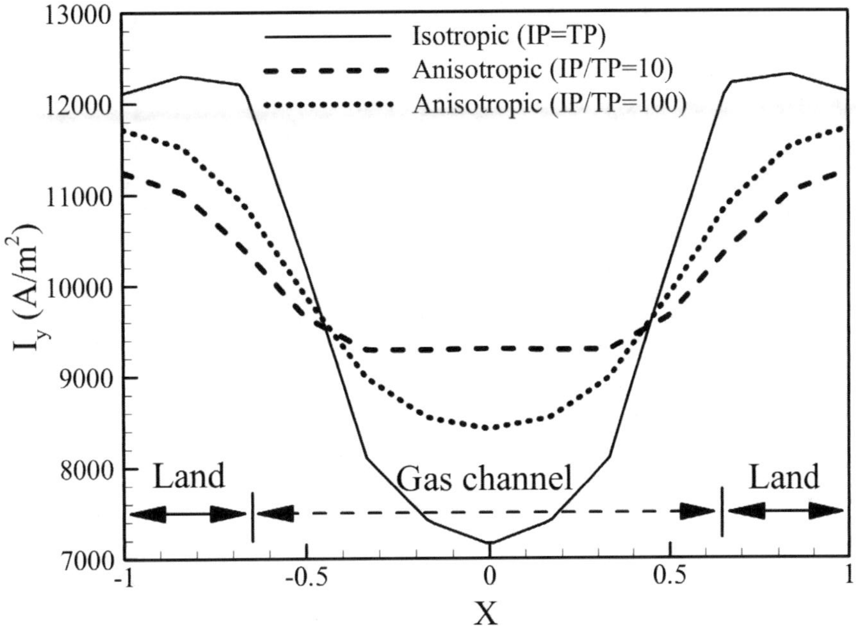

Figure 4. Predicted profiles of through-plane current on cathode GDL/catalyst layer interface

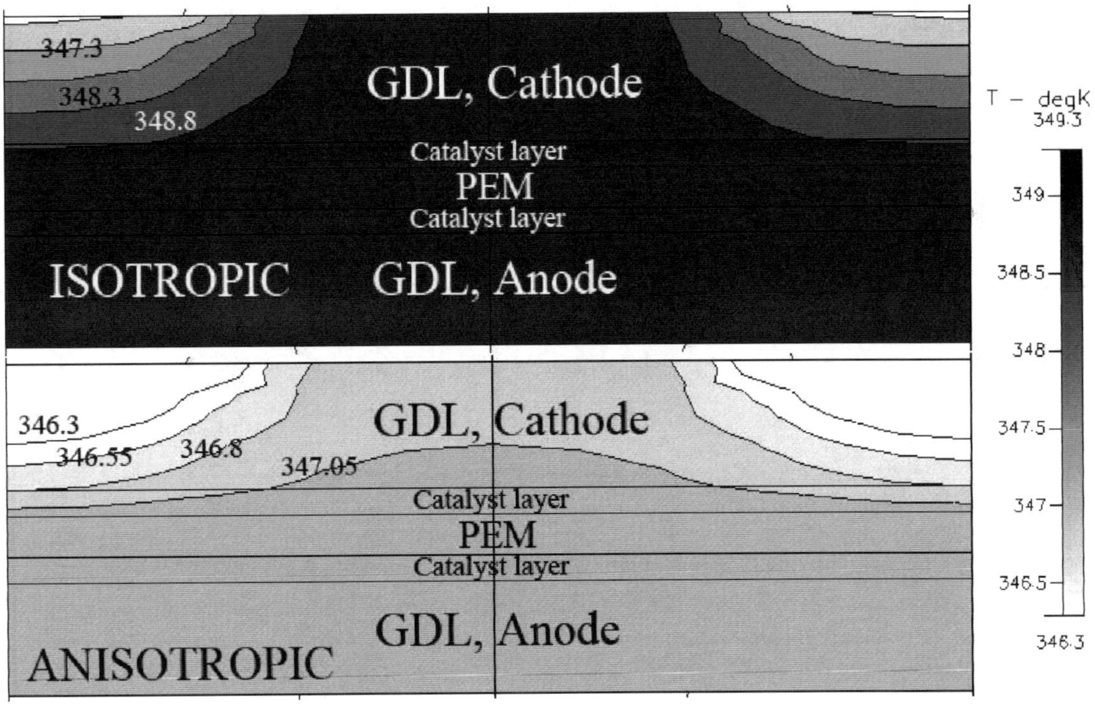

Figure 5. Predicted temperature of the MEA

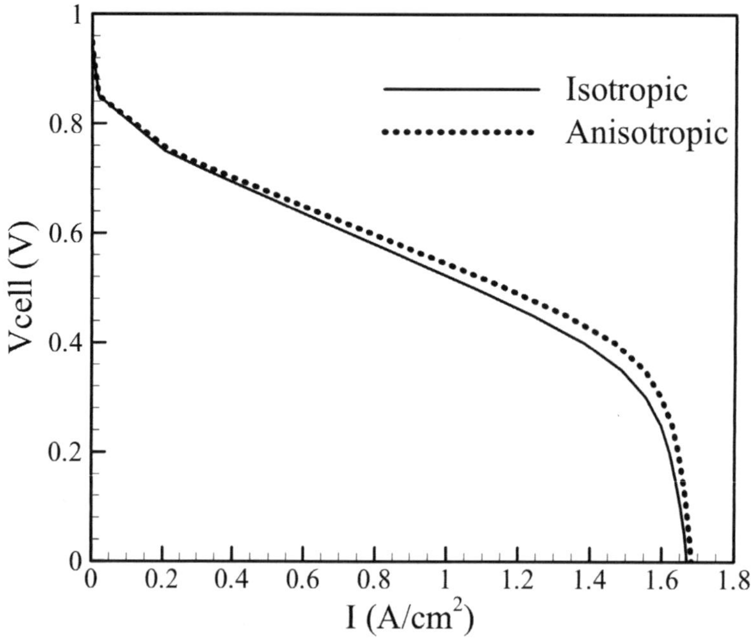

Figure 6. Polarization curve predictions

References

[1] W.-k. Lee, J.W. Van Zee, A. Jena and K. Gupta: Fuel Cell Seminar, 2004.

[2] S. Mazumder and J. V. Cole: J Electrochem. Soc., 150(11), pp. A1503-A1509, 2003.

[3] J.H. Nam and M. Kaviany: Int. J. of Heat and Mass Transfer, 46(24), pp. 4595-4611, 2003.

[4] J.G. Pharoah: J. Power Sources, 144(1), pp. 77-82, 2005

[5] J. H. Stumper and H. Haas, et al.: J. Electrochem. Soc., 152(4), pp. A837-A844, 2005.

[6] P.C. Sui and N. Djilali: Conference of Metallurgists, Calgary, AB, Canada, 2005.

[7] P.C. Sui, S. Kumar and N. Djilali: To appear at the 4th International ASME Conference on Fuel Cell Science, Engineering and Technology, June 19-21, 2006, Irvine, CA, U.S.A.

[8] W. Sun, B.A. Peppley and K. Karan: J. Power Sources, 144 (1), 42-53, 2005.

The compressed non-uniformity effect of gas diffusion layer on PEMFC performance

P. H. Chi*[a], S. H. Chan[a], F. B. Weng[a], Ay Su[a] and C.Y. Lin[b]

[a]Department of Mechanical Engineering & Fuel Cell Research Center, Yuan Ze University, Chung-Li, Taiwan, R.O.C.

[b]Department of Mechanical Engineering, Nanya Institute of Technology

ABSTRACT

The purpose of the present study is to investigate both experimentally and theoretically the effect of GDL porosity non-uniformity due to clamping force on fuel cell performance. In the experimental study, a unit cell with a single serpentine channel is employed to test the effect of compression force on cell performance. The degree of GDL deformation is achieved by varying the thickness of gasket spacer. In the numerical simulations, a three dimensional model is developed to simulate coupled electrochemical kinetics, current distribution, hydrodynamics, and multi-component transport. The properties of the GDL used in the simulation are expressed as functions of the compression ratio. The simulation results are found to be in good agreement with experimental data in overall fuel cell performance. The distributions of temperature, heat flux, species concentration, current density and saturation are found to be highly oscillating in nature between the local rib and channel locations. Furthermore, the higher the compression ratio, the better is the cell performance and the larger is the fluctuation amplitude. Finally, the higher the compression ratio, the more are the saturation, water flooding and hydrogen deficiency downstream. More detail compression effects on membrane conductivity, etc, are also presented.

Keywords: Gas diffusion layer, deformation, numerical simulation, fuel cells

---

* Corresponding author

Phone: +886-3-4618691, FAX: +886-3-4555574, Email: glixlink@msn.com

1. INTRODUCTION

Performance of proton exchange membrane (PEM) fuel cells is impacted by the properties of the gas diffusion layer which can affect the optimum performance of the catalyst and electrode and, therefore, plays an important role in cell performance [1]. The gas diffusion layer (GDL) used in PEM fuel cells is made of either carbon fiber paper or carbon cloth. Ideally, the GDL has the function of efficiently transporting reactant gases to catalyst layers, removing liquid water to the gas channels, conducting electrons with low resistance, and maintaining low contact resistance at contact interfaces. Such mass transport occurring in the diffusion layer has been shown to have a strong effect on PEM fuel cell performance [2]. In a fuel cell, or stack, the cell components are held together under a high compressive load to prevent gas leakages. Gas leakages cause poor performance and lead to potentially dangerous situations; while over-compressing the GDL increases mass transfer resistance and thus reduces cell performance. High installation pressure can generally decrease the contact resistance and it has been shown that physical compression of the GDL directly increases its electrical conductivity [3]. Most of the studies in this area have assumed, for simplicity, constant porosity of the GDL or, in other words, a uniformly compressed GDL. However, this cannot reflect the important effects on fuel cell performance of variation in GDL porosity, which becomes non-uniform under compression. The following section provides a brief review of related research. Section 2 provides the specifics of the experiment and model development and the results and discussion can be found in section 3 along with recommendations for further research.

Several recent studies have begun to explore the effect on performance of compression on the GDL. Zhou et al. [4] used a two-dimensional model of the rib and GDL to investigate the effect of clamping force on the interfacial contact resistance and the porosity of the gas diffusion layer (GDL) in a proton exchange membrane fuel cell (PEMFC). The numerical results showed that a larger clamping force and wider rib lead to smaller contact resistance. However, the porosity of the GDL was observed to decrease with increasing clamping force. Rectangular-shaped ribs were found to lower the contact resistance and porosity of GDL to a greater degree than semicircular-shaped ribs which was improved fuel cell performance. Lee et al. [5] experimentally investigated cell performance as a function of compression pressure with different commercial types of gas diffusion layers. Their investigation showed that changes in cell performance may be due to changes in the porosity, the contact resistance of the gas diffusion layer, and the excluded water in the membrane. Further, optimum cell performance was found to depend on the gasket thickness and the degree of compression of the diffusion layer. Ge et al. [6] studied the effects of GDL compression on fuel cell performance for different compression ratios (operating thickness / original thickness) and under different fuel cell operating conditions. They found that an optimal compression ratio exists and that the ratio occurs at the same percentage for different running conditions. However, the value of this ratio was found to change when different types of GDL (such as carbon fiber paper or carbon cloth) were used. Roshandel et al. [7] considered the spatial variation in porosity due to compression pressure and water produced effects. They postulated that a sine wave type porosity distribution exists underneath the landing area (rib) of GDL when a clamping force is applied. Chu et al. [8] used a one-dimensional half-cell model to investigate the effects of non-uniform porosity on fuel cell performance in terms of physical parameters such as oxygen consumption, current density, power density, etc. The non-uniformity of porosity was accounted for by four different continuous functions of position, each of which had a different averaged value of porosity and a different type of distribution across the diffusion layer. Their results showed that fuel cell performance deteriorates significantly due to a lower porosity effect in the GDL as the cathode is flooded with water.

In both of the latter studies (Roshandel et al. [7] and Chu et al. [8]) variable porosity was accounted for, but constant permeability of the GDL was assumed. In recent years, however, the assumption of constant permeability has come under scrutiny and new studies have appeared that tackle the issue of how changes in the permeability of the gas diffusion layer may impact PEM fuel cell performance. Williams et al. [1] experimentally observed that over-land convection could significantly improve cell performance and delay the onset of the mass transfer limiting regime. Ihonen et al. [9] reported that the use of high clamping pressure decreased the permeability of the GDL. J.G. Pharoah [10] used computational fluid dynamics (CFD) to determine the importance of convective transport and how it operates as a function of GDL permeability when beyond a threshold permeability value of approximately $1 \times 10^{-13} m^2$, assuming constant porosity in the GDL.

However, in their recent experimental study, Gostick et al. [11] found that if a GDL is compressed to the half of the initial thickness, its permeability is decreased by an order of magnitude. Thus, all previous research suggests that the compression of the diffusion layer will generally influence not only the conductivity and contact resistance but also the porosity and permeability of the diffusion layer and affect cell water management, as well.

However, compression force is not uniform across the GDL layer due to the rib and channel design of PEM fuel cells. When clamping force is applied, the porosity of areas of the GDL in contact with the ribs of the bipolar plate can be significantly different from areas of the GDL in contact with the channels of the bipolar plate. This difference can give rise to local variation phenomena which can affect the built-in design durability of PEM fuel cells, an important concern in addition to the overall performance of the fuel cell. For instance, a non-uniform temperature distribution can cause thermal stress in the membrane and catalyst layers and a non-uniform saturation distribution can cause local water flooding which can reduce the useful reactive area. Therefore, it is also important to model the effects of non-uniform compression force to fully understand the processes at work in the fuel cell and their implications for performance.

While the text of research discussed in the preceding paragraphs has helped us to understand the separate effects of GDL compression on porosity and permeability and corresponding effects on fuel cell performance, to date no prior numerical studies have considered the effects of variation in both the permeability and porosity of GDL compression, simultaneously. Similarly, no extant studies have utilized a non-uniform model of GDL compression weighed against experimental data to explore which of the models fits best with the empirical data. Thus, the aim of the present study is to investigate the effects of simultaneous variation of permeability and porosity due to GDL compression as well as non-uniform compression model in explaining the experimental results and providing important data about fuel cell processes in GDL compression situations.

More specifically, this study investigates this non-uniform porosity effect for a non-uniform square wave variation model allowing for a smaller porosity under the rib (the compressed part) and for a larger porosity under the flow channel (the uncompressed part). A three dimension numerical simulation analysis is adopted here to better understand local phenomena, such as temperature, current density, species, saturation (liquid water), and membrane conductivity distributions, in addition to the overall performance of PEM fuel cell. Thus, the specific objectives of this work are to (1) to validate the detailed 3-D numerical PEM fuel cell model through the comparison of fuel performance with experimental results under different compressed ratios of GDL, and (2) to study local phenomena in detail to help explain how different GDL compression ratios affect fuel cell performance.

2. EXPERIMENTAL SETUP AND MODEL DEVELOPMENT

2.1 Experimental Setup

A unit cell is employed to obtain cell performance data under different GDL compression ratios. The hardware of the unit cell, from inside out, consists of a membrane electrode assembly (MEA, which includes catalyst layers, a membrane and GDLs), flow field plates, a heating plate, and end plates, see Fig. 1. Graphite plates with serpentine flow pattern are used as the current collector as well as the flow field plates. The MEA (Gore 5621, GoreTex Fuel Cell Co.) has a reaction area 25 cm^2. For each compression ratio case, a Teflon gasket of known thickness is placed around the active area of the MEA as a spacer to control the deformation of the MEA. Three gaskets with the thickness of 0.256, 0.207 and 0.158 mm respectively are used on both the anode and the cathode sides in the experiments. In the present study the compression ratio (CR) is defined as the ratio of the reduced thickness versus the original thickness,

$$CR = \frac{L - l}{L}, \tag{1}$$

which differs from other researchers, e.g. [12]. The CR value by our definition is proportional to the clamping force and hence the compression pressure. The original thickness (L) of GDL is 0.4mm. Assuming the gasket is the much stiffer than the MEA, the thickness (l) of the compressed GDL is the gasket thickness. The CR values for the 0.256, 0.207 and 0.158 mm gaskets are 36%, 48.3% and 60.5%, respectively.

The dimensions of components and operating conditions are set at Table 1.

A fuel cell test system (ScribnerAssociates, model 850C) is employed to record polarization curves and the high frequency ohm resistance/voltage curve of the test cell. Hydrogen and oxygen at 300 c.c./min. and fully-humidified at 65°C are used as the fuel and oxidant respectively. The cell temperature is maintained at 70°C. The outlet pressure of the anode and the cathode is 1 atm. The anode and cathode flow each enters the unit cell from the top in a counter-flow manner, see Fig. 2.

2.2 Model Development

Most of the literature on CFD model for PEMFC considers GDL of uniform properties, e.g. Djilali[13], Wang[14], etc. Recently Su et al. [14,15] reported a 3D CFD model that considered non-uniform property distributions for the GDL, although the GDL domain did not take into consideration the actual deformation of the material. In the present study the non-uniformly compressed GDL, cf. Fig. 1, which contains compressed parts and uncompressed parts is considered. The porosity of GDL underneath the ribs is decreased by the compression force. Three CR values (36%, 48.3% and 60%) are used in the experiments and numerical simulation for comparison. An addition CR value of 70% is included in the numerical simulation to predict the cell performance under extreme compression conditions.

2.2.1 Assumptions for the Numerical Model

The numerical simulations performed in the present study are based on the solution of the conservation equations of mass, momentum, energy, saturation, current and species transport on a computational grid using finite-volume methodology. The key elements in the modelling of fuel cells are the transport phenomena through porous media, heterogeneous reactions within porous electrodes and the coupling between mass transports, electrochemical reactions and current–potential fields. A commercial CFD software [16], CFD-ACE+ v. 2003, is employed in the present study. The theoretical background and validation of this simulation tool can be found in recent papers by Sui et al[15]. For the 3D simulations performed in the present study, we make the following assumptions:

1. The system reaches a steady state.
2. Laminar flow and ideal gas for the fluids considered.
3. Stefan-Maxwell for multi-component gas diffusion.
4. The electrochemical kinetics is described by the Butler-Volmer equation.
5. Proton transport in the ionomer is described by the Nernst-Planck equation.
6. Negligible contact resistance.
7. The deformation of GDL is elastic.

According to the assumptions, the governing equations corresponding to the various regions of the fuel cell are given below [17]. The simulation code itself has been verified elsewhere [18].

2.2.2 Porous media properties

The thermal conductivity λ is a combination of the porosity, solid and fluid thermal conductivity and is defined by following equation [19, 20],

$$\lambda = -2\lambda_s + \frac{1}{\dfrac{\varepsilon}{2\lambda_S + \lambda_F} + \dfrac{1-\varepsilon}{3\lambda_S}} \tag{2}$$

where the λ_S and λ_F are the solid and fluid thermal conductivity, ε is the porosity of the porous media. When $\lambda_S >> \lambda_F$, this reduces to

$$\lambda \sim \frac{2\lambda_S(1+\varepsilon)}{2+\varepsilon} \tag{3}$$

The species diffusivity in the GDL is related to porosity using the Bruggeman relation [20, 21]:

$$D_i = D_{i,FS} \cdot \varepsilon^{\tau} \tag{4}$$

where τ is the tortuosity factor and $D_{i,FS}$ is the free space diffusivity.

2.2.3 Compressed GDL Model

The effect of compression on GDL is assumed to be elastic and the clamping force does not destroy the solid fibbers of the porous media but merely reduces the pores space. The edge effect is neglected for its relatively smaller scale. The porosity of the compressed GDL is closely related to the pore space, which can be readily derived to yield the following equation:

$$\varepsilon_{compressed} = \frac{\varepsilon_{original} - CR}{1 - CR} \tag{5}$$

The absolute permeability (κ) is a function of porosity (ε) and is defined by the Carman-Kozeny equation [22, 23]:

$$\kappa = \frac{D_r^2 \cdot \varepsilon^3}{16 \cdot C_k (1-\varepsilon)^2} \tag{6}$$

where the Kozeny constant $C_k = 6$ and particle diameter $D_r = 7 \times 10^{-6}$.

2.2.4 Computational Domain and Boundary Conditions

The 3D CFD computational model resolves most of the components used in the experiments, i.e. GDLs, catalyst layers, membrane, gas channels and solid bipolar plates. The computational domains are divided into 32 elements in the *X*-direction, 164 elements in the *Y*-direction, and 49 elements in the *Z*-direction. The total number of the simulation model contains 257,152 cells. The grid distribution is found to be adequate for reproduction of simulation results with higher number of cells. At gas inlets, gas composition and mass flow rate is prescribed and at gas outlets the pressure is set to be 1 atm. The temperature boundary condition on both sides of the current collector is set to be constant while the remaining surface of the domain is set to be adiabatic.

3. RESULTS AND DISCUSSION

The properties, parameters and baseline conditions used in the numerical simulations are listed in Table 1 and 2. Three GDL compression cases (CR=36%, 48% and 60%) and parameters for the experiments and numerical simulations are shown in Table 3.

In the follow discussion of simulation results, the distribution of temperature, saturation and current density distributions are presented on a Z-cut plane, cf. Fig.2, through the cathode catalyst layer (CCL), while the membrane electrical conductivity distribution are presented on a Z-cut plane through the membrane layer. The species concentration is presented through the CCL for oxygen and the ACL for hydrogen. In the mean time, we cut the Z-plane (such as CCL, PEM and ACL) and the centre X-plane to yield a Y-Y line along the Y direction, cf. Fig. 2. The data shown along the Y-Y line reveal the variation from the inlet up-stream all the way to the outlet down-stream location. The channel and rib widths are both 0.0012 m. The inlet channel width is located at y axis point 0.0468 m to 0.048 m. The outlet channel width is position at y axis point 0 m to 0.0012 m.

3.1 Comparison of Cell Performance of Simulation and Experiment Data

The experimental data of cell performance and high frequency resistance are shown in Fig. 3. The resistance decreases with increasing compression ratio and the cell performance increases with increasing compression ratio. However, the I-V curves also show that the concentration polarization is affected by CR values, namely, concentration polarization appears at higher CR cases. This is because at larger CR value, the porosity under the rib tends to decrease; therefore mass transfer limitation is likely resulted therein.

The predicted polarization curves for four CR cases (Sim-36, Sim-48, Sim-60 and Sim-70) are shown in Fig. 4, and are compared with the experimental results at similar conditions except the Sim-70 case of which experimental data are not available. The general trend of predictions agrees well with that seen from experimental data and the agreement appears to be better for the cases with higher CR values. It is noted that the simulated I-V curves extend slightly furthers into the higher current density range such that the simulated concentration polarization begins at the lower voltage value than the experimental value. This can be explained as a result of the step function of non-uniform porosity variation used in the simulation. In the numerical model, the porosity variation is assumed as a step function along the rib-channel direction, namely, only the portion of the GDL immediately under the rib is assumed compressed, whereas in reality the compressed region may extend slightly beyond the edge of the rib. Thus the slightly higher porosity allows for more mass transfer and delays the onset of concentration polarization to a higher current density condition.

3.2 Comparison of different compressed ratio effects

On the basis of the present non-uniform model, effects of different compression ratios (CR) are examined. From the previous IV curves of Fig. 4, it is seen that the higher the CR, the better is the performance in the working voltage range and the earlier is the occurrence of concentration polarization at higher voltages. This is because the higher compression induces less contact resistance, shorter GDL thickness and less pores of GDL to reduce the high frequency resistance and ohmic polarization. Although the less pores of GDL can increase the electric transfer cross section area, it can also reduce the mass transfer of reactants (fuel or oxidant) to or the product (water vapour) released from the catalyst layer. In consequence, the higher CR value reduces the high frequency resistance to improve the performance in the cell working voltage range (0.4~0.65V), and also causes the earlier concentration polarization below the cell working voltage range.

Figure 5 and 6 show the respective normalized distributions from the upstream inlet to the downstream exit along the Y-Y line with different compressed ratio of 36% (Sim-36), and 60% (Sim-60). Substantial difference exists. For 36% CR(E-36N in fig.5), current density is the lowest but stable, saturation rises from upstream to over 0.2 after rib 13 and then to the maximum value of 0.5 at the downstream exit, membrane conductivity rises up to the maximum after rib 12, and some hydrogen mass fraction remains at exit. For 60% CR, Fig.6 shows that current density is higher than 36% CR value and drops, rather than stable, after rib 10, saturation rises faster and is over 0.2 after rib 9 and 0.6 maximum at exit, membrane conductivity also rises faster to the maximum at rib 10 , and little hydrogen mass fraction remains after channel 15. For even higher CR value of 70%, Fig.5 shows that current density is the highest but drops faster at the further upstream location of rib 9, saturation rises even faster and reaches the maximum of 0.8 at exit, membrane conductivity increases to the maximum after rib 8, and hydrogen mass fraction depleted earlier after channel 13. Thus several conclusions can be reached. (1) The higher the compression ratio, the higher and the faster is the saturation rise along the stream. The finial saturation value at the exit is proportional to the CR value. (2) Similarly, the current density is larger for higher compressed GDL, but drops down at the earlier downstream location due to flooding (saturation). (3) The earlier flooding causes the earlier hydrogen deficiency along the stream. (4) Finally, the electrical conductivity of membrane also increases faster with higher compression.

A more detail direct comparison between the low 36% and high 70% compression ratio effects is shown in Fig.7. Figure 7(a) shows that 36% CR (Sim-36) yields a fairly uniform temperature distribution, but the high 70% CR(Sim-70) in Fig.7(e) produces not only large local temperature oscillation amplitude between neighboring channel and rib locations but also even larger variation between the upstream and downstream locations. Figures 7(b) and (f) compare the saturation distribution in CCL. They clearly show that high CR produces much more liquid water. Figures 7(c) and (g) show that low CR yields an overall uniform current density distribution while the high CR produces a higher upstream but a lower downstream current density distribution due to water flooding. Comparison between Fig.7(d) and (h) shows that hydrogen mass fraction depleted only near the exit in case of low CR, but almost a third of area is depleted and therefore useless in high CR case due to water flooding. In short, the higher the CR, the more saturation occurs at downstream and the less are the temperature, current density and hydrogen at downstream.

4. CONCLUSIONS

The conventional numerical simulation model of PEM fuel cells assumes that the gas diffusion layer (GDL) can be treated as a uniformly compressed layer. The present study argues that a more realistic model of the GDL is as a non-uniformly compressed layer because the installation force compresses mainly on the ribs against the GDL. For each parameter, the performance predicated by each model was evaluated against the experimental data to assess which model fit better with the results. The following major conclusions are made:

The distributions of temperature, heat flux, species concentration, current density and saturation are found to be highly oscillating locally between the rib and channel locations. The higher the compression ratio, the larger is the fluctuation amplitude.

The temperature and hydrogen distributions are fluctuating in the same phase, higher in channel and lower in rib locations. The saturation, heat flux, current density and membrane conductivity also fluctuate in phase but to a lesser degree in fuel cell channels and to a higher degree in rib locations. The saturation fluctuating distribution reveals that the temperature is lower beneath the rib region which causes local condensation and higher saturation. This phenomenon increases membrane conductivity and cell performance. At the same time, however, it can cause water flooding and decrease the cell performance downstream.

The larger the CR, the greater the saturation occurring downstream and the lower the temperature, current density and hydrogen concentration downstream. The larger the CR value (for the range investigated) the higher is the overall total performance, but the resultant water flooding downstream leads to hydrogen deficiency and less effective use of reaction area downstream. Higher CR values yield greater local temperature oscillation and larger overall temperature variation which may impose higher local thermal stresses on membrane and catalyst layers.

REFERENCES

[1] M.V. Williams, E. Begg, L. Bonville, H. Russell Kunz, J.M. Fenton, J. Electrochem. Soc. 151 (8) (2004) A1173–A1180.

[2] B. Thoben, A. Siebke, J. New Mater. Electrochem. Syst. 7 (2004) 12–20.

[3] J. Nordlund, *The anode in the direct methanol fuel cell*, Doctoral Dissertation.

[4] P. Zhou, C.W. Wu, G.J. Ma, J. Power Sources 159 (2006) 1115–1122.

[5] W.K. Lee, C.H. Ho, J.W. Van Zee, M. Murthy, J. Power Sources 84 (1999) 45–51.

[6] Jiabin Ge, Andrew Higier, Hongtan Liu, J. Power Sources 159 (2006) 922–927.

[7] R. Roshandel, B. Farhanieh, E. Saievar-Iranizad, Renewable Energy 30 (2005) 1557-1572.

[8] Hsin-Sen Chu, Chung Yeh, Falin Chen, J. Power Sources 123 (2003) 1-9.

[9] Jari Ihonen, Mikko Mikkola, and Göran Lindbergh, J. Electrochem. Soc., 151 (8) A1152-A1161 (2004)

[10] J.G. Pharoah, J. Power Sources 144 (2005) 77-82

[11] Jeff T. Gostick, Michael W. Fowler, Mark D. Pritzker, Marios A. Ionnidis, Leya M. Behra, J. Power Sources 162 (2006) 228-238

[12] Fang-Bor Weng, Ay Su, Yur-Tsai Lin, Guo-Bin Jung, and Yen-Ming Chen, J. of Fuel Cell Science and Technology, Vol. 2, I 3 (2005), pp. 197-201.

[13] Y.M. Ferng,A. Su, Int. J. of Hydrogen Energy, 32 (2007), 4466-4476.

[14] P.K. Sinha, C. Y. Wang, and A. Su, Int. J. of Hydrogen Energy, 32 (2007) 886-894.

[15] P.C. Sui, N. Djilali, J. Power Sources 161 (2006), 294-300

[16] CFD-ACE+ user maanual.

[17] Sandip Mazumder and James Vernon Cole, J. Electrochem. Soc. 150(11), (2003) A1503-1509.

[18] P.H. Chi, Fang-Bor Weng, Ay Su, and S.H. Chan, J. of Fuel Cell Science and Technology, Vol. 3, I 3 (2006), pp. 225-365.

[19] V. Gurau, H. Liu, and S. Kakac, AIChE J., 44 (1998), pp. 2410-2422.

[20] Dagan, G., 1989, *Flow and Transport in Porous Formations*, Springer, Berlin

[21] Kaviany, M., 1995, *Principles of Heat Transfer in Porous Media*, second ed., Springer, New York

[22] Bird, R. B., Stewart, W. E., and Lightfoot, E. N., 1960, *Transport Phenomena*, Wiley, New York.

[23] M.M. Tomadakis, T.J. Robertson, J. Compos. Mater. 39 (2005) 163–188.

Notation

CR	:	compression ratio
δ	:	thickness of compressed GDL
L	:	thickness of uncompressed GDL
l	:	spacer gasket thickness
ε	:	porosity
$\varepsilon_{orinigal}$	:	porosity of uncompressed GDL
$\varepsilon_{compressed}$	:	porosity of compressed GDL
κ	:	absolute permeability (m^2)
λ	:	thermal conductivity
λ_S	:	solid thermal conductivity
λ_F		fluid thermal conductivity
τ	:	tortuosity
$D_{i,FS}$	:	the free space diffusivity

Table 1. Base operating conditions for the numerical model & experiment

Channel width	1.2 mm
Rib width	1.2 mm
Channel height	1 mm
GDL thickness	0.4 mm
Catalyst layer thickness	Anode= 0.018 mm, Cathode= 0.026 mm
Membrane thickness	0.035mm
Total reaction area	25 cm^2
Effective diffusivity	Bruggeman,τ=7 for membrane Bruggeman,τ=1.5 for GDL and catalyst layer
Membrane permeability	1.8×10^{-18} m^2
Catalyst layer permeability	1.76×10^{-11} m^2
Membrane porosity	0.28
Catalyst layer porosity	0.4
Transfer coefficient (Tafel constants) at anode	0.5
Concentration dependence at anode	0.5(H_2)
Reference current density at anode	1.0×10^{9} (A/m^3)
Transfer coefficients (Tafel constants) at cathode	1.5
Concentration dependence at cathode	1.0 (O_2)
Reference current density at cathode	1.5×10^{3} (A/m^3)
GDL and catalyst layer conductivity	100 1/Ω m

Table 2. Different cases and corresponding parameters for the numerical model & experiment

	Anode	Cathode
Inlet gases	H_2	O_2
Inlet volume flow rate(cc/min)	300	300
Inlet gases relative humidity (%)	100	100
Inlet temperature(°C)	65	65
Operating pressure(atm)	1	1
Cell temperature(°C)	70	70

Table 3 Dimensions, properties and parameters for the numerical model

Case	Gasket thickness (mm)	Compression Ratio (CR, %)	Uncompressed Porosity	Compressed Porosity	Permeability $\times 10^{-12}$
Sim-36	0.256	36.0	0.8	0.69	3.198
Sim-48	0.207	48.3	0.8	0.61	2.091
Sim-60	0.158	60.5	0.8	0.49	1.108
Sim-70	0.120	70.0	0.8**	0.33	0.495
Exp-36	0.256	36.0			
Exp-48	0.207	48.3			
Exp-60	0.158	60.5			

Original GDL thickness =0.4 mm

Sim-xx : Numerical simulation cases with different compressed ratio.

Exp-xx : Experimental cases with different compressed ratio.

** : Permeability of 0.8 porosity is 6.533×10^{-12}

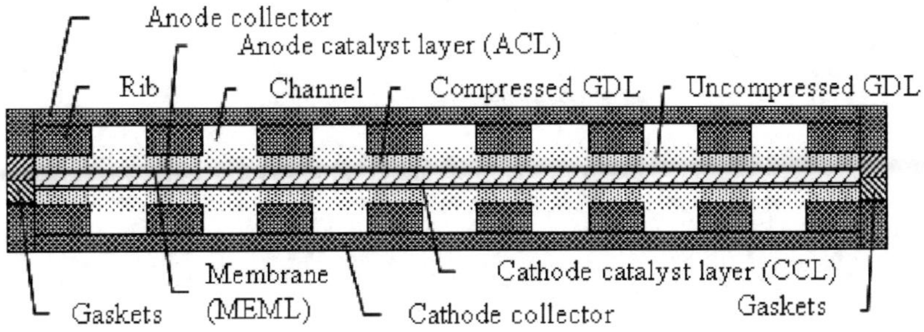

Figure 1. Schematic illustration shows the compressed GDL of PEMFCs

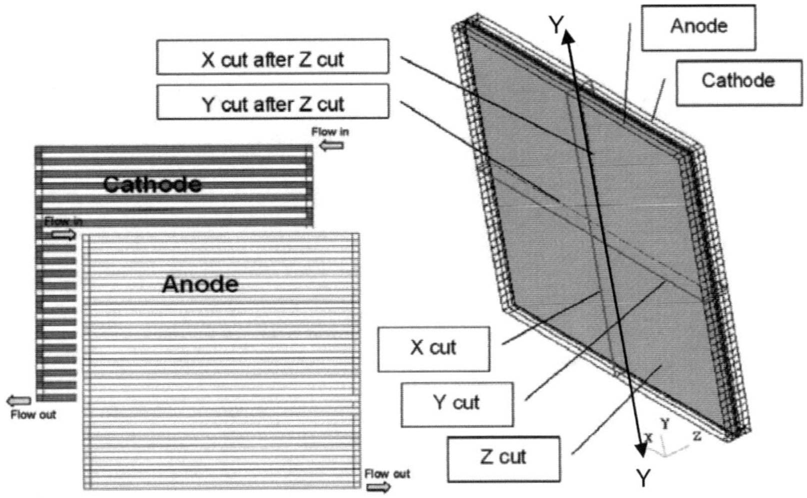

Figure 2 The Z cut plane, X cut plane and line of X cut after Z cut (Y-Y line) are focus domains; and the anode/cathode flow direction

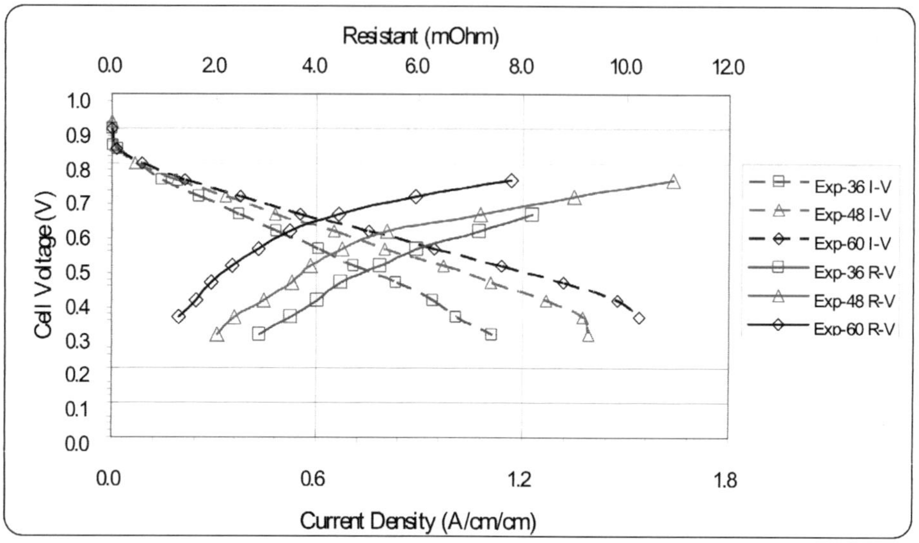

Fig. 3 Experimental data for I-V curve and R-V curve

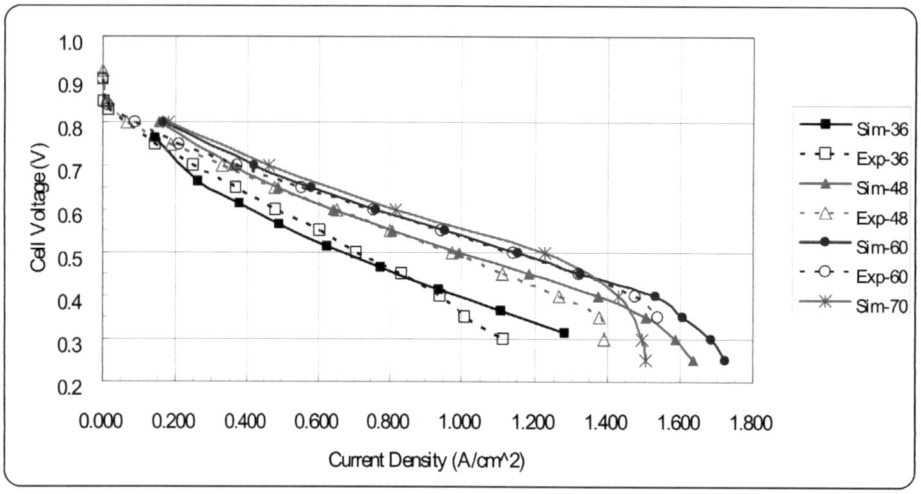

Fig. 4 Polarization curves for different CR value of experiment and simulation data

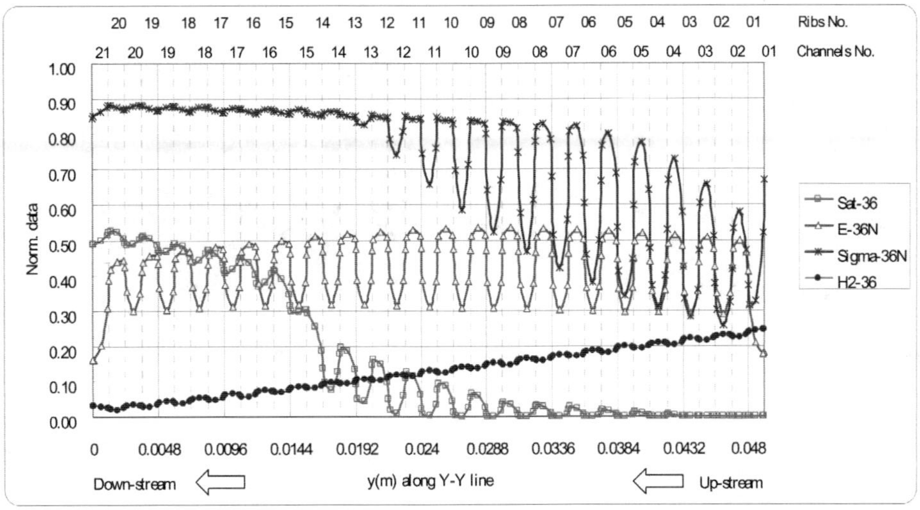

Fig. 5 Normalize data along Y-Y line for different compressed ratio of GDL CR=36% (Sim-36) at operation voltage 0.3V.

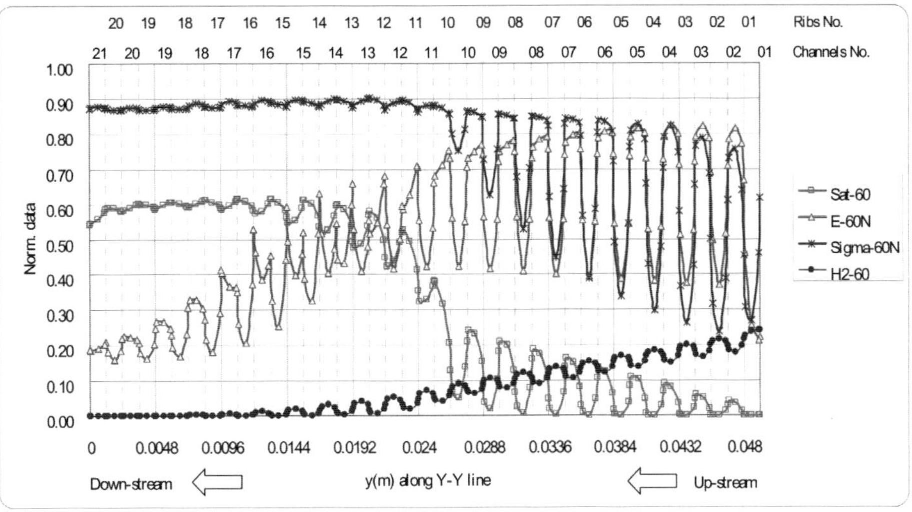

Fig. 6 Normalize data along Y-Y line for different compressed ratio of GDL CR=60% (Sim-60) at operation voltage 0.3V.

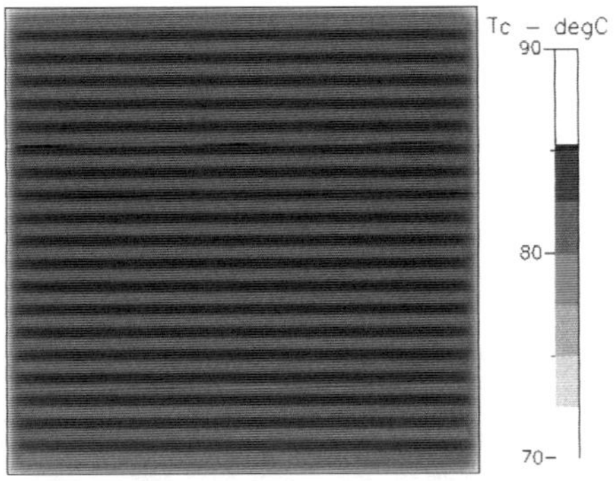

(a) Temperature of Sim-36 at ccl

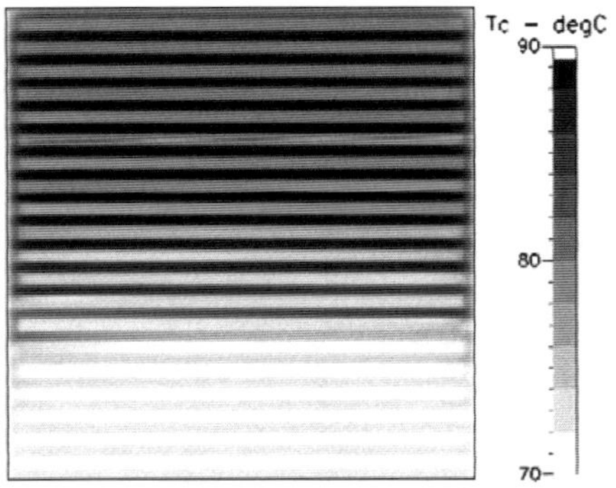

(e) Temperature of Sim-70 at ccl

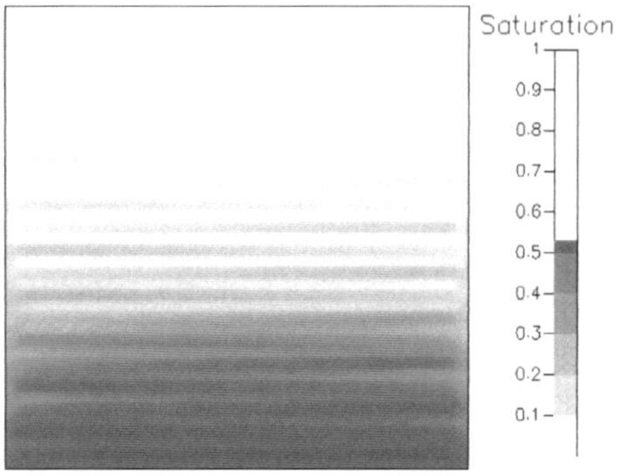

(b) Saturation of Sim-36 at ccl

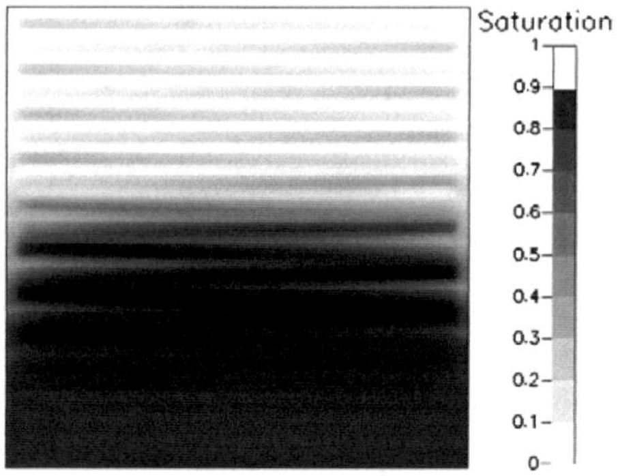

(f) Saturation of Sim-70 at ccl

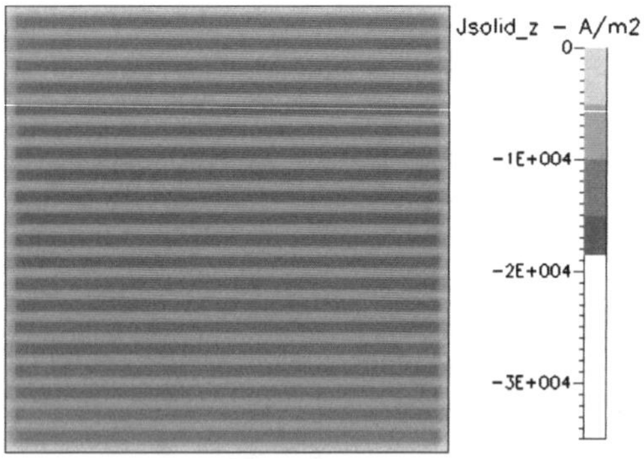

(c) Current density of Sim-36 at ccl

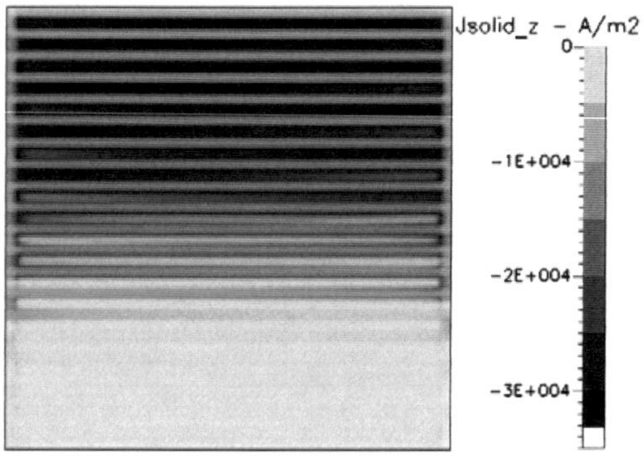

(g) Current density of Sim-70 at ccl

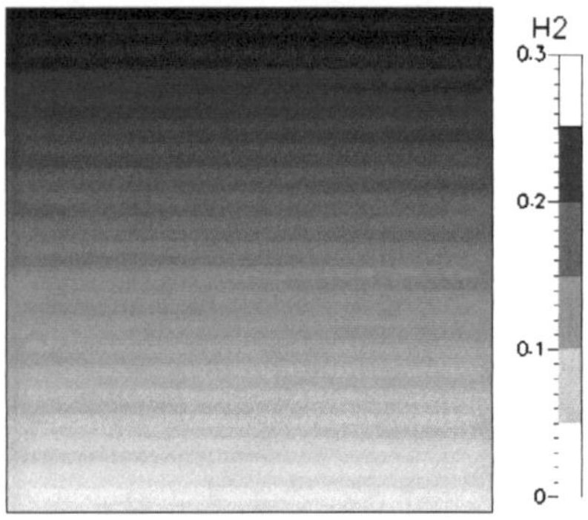

(d) Hydrogen of Sim-36 at acl

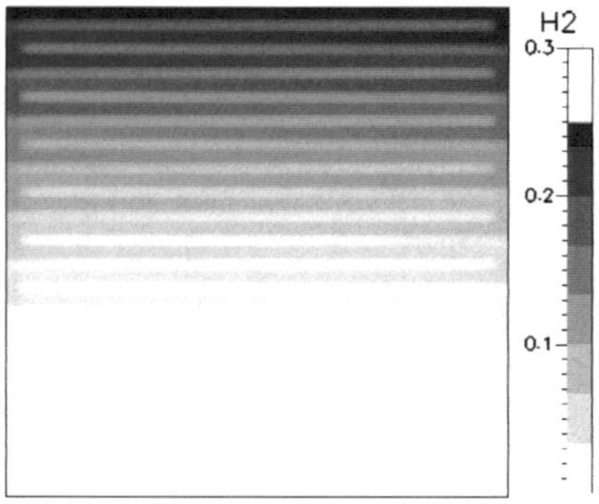

(h) Hydrogen of Sim-70 at acl

Fig. 7 Compare the saturation, temperature, current density and hydrogen mass fraction of Sim-36 ((a1), (b1), (c1) and (d1)) and Sim-70 ((a2), (b2), (c2) and (d2)) at cell voltage of 0.3V.

538e Development of Cathode Air Filters for PEM Fuel Cell Using Microfibrous Entrapped Sorbents

Abhijeet G. Phalle[1], Vivekanand Gaur[1], and Bruce Tatarchuk[2]. (1) Department of Chemical Engineering, Auburn University, 230 Ross Hall, Auburn, AL 36849, (2) Chemical Engineering, Auburn University, 230 Ross Hall, Auburn, AL 36849

The performance of Proton Exchange Membrane Fuel Cell (PEMFC) is degraded significantly as a result of poisoning of cathode catalyst by airborne contaminants, in particular SO_2 and NO_x in trace levels. In the literature, it has been suggested that the best method for addressing fuel cell contamination is by the inclusion of adsorptive filtration with a cathode air filter. This work thus aims at the development of adsorptive cathode air filters, mainly to remove SO_2 because of its irreversible effects on PEMFC. Two types of sorbents were prepared; an alumina support impregnated with different transition metal (Mn, Fe, Ni, Cu and Zn) oxides and inorganic chemicals (KOH and $KMnO_4$) at varied loadings. The adsorption capacity of these sorbents was investigated in a tubular packed column with inlet SO_2 concentration of 70 ppm in air under dry as well as moist (50% relative humidity) conditions at room temperature. At identical loadings,among all metal oxide based sorbents, MnO_2/Al_2O_3 showed the highest saturation capacity. The SO_2 adsorption capacity of the screened sorbents at saturation level decreased in a sequence as: $KMnO_4$ (5%)/Al_2O_3 > MnO_2 (10%)/Al_2O_3 > KOH (10%)/Al_2O_3. In contrary, the results demonstrated that the 10% (w/w) KOH/ Al_2O_3 had almost double breakthrough capacity for SO_2 as compared with that of 10% (w/w) MnO_2/Al_2O_3.

A bi-component polymer fibers-entrapped KOH/Al_2O_3 sorbent (PFES) is designed for SO_2 removal from air. Due to the microfibrous media and supported sorbent design, PFES demonstrated superior SO_2 removal performance as compared with a packed bed. At equivalent (10 wt%) KOH loading, the PFES yielded 50% higher breakthrough capacity than that for KOH/Al_2O_3 sorbent in the packed bed. This is attributed to the high contact efficiency of small particulates, high KOH utilization, and high accessibility of KOH. The other sorbents including $KMnO_4/Al_2O_3$ and MnO_2/Al_2O_3 are also being entrapped in microfibrous sheets and investigated for their SO_2 removal efficiency. The use of these novel microfibrous entrapped sorbents provides reduced external mass transfer resistance enhancing adsorption rate. Similar entrapped sorbents will be developed for the air contaminants other than SO_2 including NO_x and VOCs.

538f A Predictive Pressure Drop Model for a Designing and Optimizing Cathode Air Filters

Ryan A. Sothen and Bruce Tatarchuk. Chemical Engineering, Auburn University, 0322A Haley Center, Auburn, AL 36849

Proton Exchange Membrane (PEM) fuel cells are susceptible to damaging airborne molecular contaminants (AMCs) such as carbon monoxide, sulfur compounds, and volatile organic compounds (VOCs). Previous research indicts that the most effective method to deal with the presence of these AMCs is to increase the effectiveness of the cathode air filter. Adsorbent entrapped media, such as microfibrous materials engineered at Auburn University, provide a novel method to efficiently entrap particles as small as ten microns within a sinter-locked matrix of fibers. The small adsorbent diameters greatly increase the contacting efficiency and performance of the media; however, the particles also increase the pressure drop due to additional drag forces. Since the PEM fuel cell's compressor is self-powered, an increase in cathode filter resistance creates a rise in the parasitic power drain on the system. Thus, the prevention efforts can become self-defeating.

Pleating the media is a simple approach to enhance the capacity while reducing the flow resistance. A pleated filter achieves this by increasing the available filtration area. In turn, this reduces the velocity into the media and pressure drop across the media. Although the additional area reduces the pressure drop across the media, the presence of pleats creates a new resistance due to fluid-solid friction. Thus, construction of a pleated becomes an optimization process of balancing pleat resistance and media resistance to obtain the lowest pressure drop.

A comprehensive study was performed to identify the contributing flow resistance factors in order to optimize a cathode air filter. By employing various fundamental fluid dynamics equations and empirical data, the research established a predictive model capable of estimating pressure drop for flow through a pleated filter. The model can be used as a design tool to identify the contributions of pleat count, filter depth, media type, and number of filtering elements on the total resistance of the system. Using the model, the capacity and contacting performance of a pleated filter is compared to a packed bed at a fixed pressure drop and volumetric flows. Through the use of novel media and packaging techniques, the amount of adsorbent present in a cathode filter unit can be dramatically increased while maintaining a lower pressure drop to reduce the parasitic effects of the filter

A Numerical Investigation on the Flow, Thermal and Electrical Coupling in a PEMFC Stack

P.C. Sui and Ned Djilali
Institute for Integrated Energy Systems, University of Victoria, Victoria, BC, Canada

Abstract

This paper reports on numerical investigations for a proton exchange membrane fuel cell stack using a CFD-based simulation tool. The stack simulation tool is constructed based on unit cells in parallel that have all components resolved and major transport equations solved in a two-dimensional configuration. The transport equations considered in the model are modified from a complete set of conservation equations (mass, momentum, energy, species and electrical potential). The stack tool also incorporates a flow network solution procedure for all gas and coolant channels to account for flow sharing of fluids at the stack level. The unit cell model in the 2D stack tool is validated with a 3D, comprehensive CFD model. The parameters used in the flow network model are calibrated from experimental data. The stack tool is used to determine the sensitivities of the baseline design of a stack with respect to design parameters, e.g. the dimensions of headers and unit cell components, and operating conditions such the oxidant stoichiometric ratio and coolant flow rate, and so on. Simulation results for the baseline stack with an anomalous unit cell (ill-behaving in flow, thermal or electrical transport) are presented to show how a stack's performance would deviate from the design point due to unfavorable conditions.

Introduction

Flow sharing and thermal management play an important role in the overall efficiency and system stability of a PEMFC fuel cell stack. In a typical plate-and-frame design of PEMFCs for automotive applications, there exist gradients of temperature, electrical potentials and gas concentration in different directions of the stack. The coupling of flow, thermal and electrical transfer in a stack is very involved, hence the design of unit cell and stack hardware is a rather complicated process. In the literature there have been several reports for stack modelling based on reduced dimensional and physical models, e.g. [1] and [2]. These models were implemented using proprietary computer codes, which were thus limited in their capabilities in application to common design practice. In order to better integrate into existing design environment, the CFD methodology is adopted for the development of a stack simulation tool. The CFD-based stack tool thus developed takes the standard procedure of preprocessing (mesh generation), solution of the governing conservation equations, and post-processing. The objective of the present study is to report on the development of the CFD-based simulation tool for the optimal design of PEMFC stacks.

Mathematical Formulation

Figure 1 shows a typical plate-and-frame type of PEMFC. A unit cell is primarily a membrane electrode assembly (MEA) sandwiched by two bipolar plates. The MEA is made of a membrane (proton conductor), coated by catalyst layers, and hot-pressed to gas diffusion layers on both sides.

There are three primary directions of transport, i.e., the direction perpendicular to the MEA, along the flow direction in the unit cell, and along the headers in the stack. These directions correspond to coordinate of z, x, and z in Fig. 1, respectively. Since the thickness of the MEA is the smallest length scale in the stack, the gradients of species and potentials are likely the largest. Along the gas/coolant channels of the unit cell, significant variation of concentration and temperature exist because of stoichiometric ratio and thermal management requirements. The flow sharing in the headers of the stack is expected to be smaller than the former two directions, however, the impact of stack level variations is tremendous because the flow and thermal distribution of the unit cells are closely coupled. To simplify the analysis, the present model makes the following assumptions: (1) The stack operation reaches a steady state; (2) The flow in the porous media and gas channels is in vapor phase; (3) Negligible contact resistance for heat and electron conduction. Because of a 2-D configuration adopted for the computational domain, further simplifications are needed: (1) The flow resistance in the fluid channels is adjusted to reflect the hydraulic diameter of the channel geometry to obtain similar pressure drop; (2) The inlet velocity of gas channels is adjusted to achieve the same stoichiometric ratio as the actual 3-D channel; (3) The thermal and electrical conductivities of the fluid channels are adjusted to allow heat flux and current density to be comparable to the actual geometry.

Governing Equations

Conservation equations of mass, momentum, energy, gas species and charged species (electron) are solved in the present study. For a steady state, the conservation equations can be expressed in the following form:

$$\rho\vec{u}\nabla\cdot\phi = \nabla\cdot\vec{J}_{\phi} + S, \tag{1}$$

with $\phi = 1, \vec{u}, h, Y_i, \Phi$ for conservation of mass, momentum, energy, gas species and charged species, respectively. The source terms for gas species and energy conservation as function of local current density are:

$$S_{Y_{O_2}} = \frac{-\nabla\cdot\vec{i}}{4F} \tag{2}$$

$$S_{Y_{H_2}} = \frac{-\nabla\cdot\vec{i}}{2F} \tag{3}$$

$$S_{Y_{H_2O}} = \frac{\nabla\cdot\vec{i}}{2F} \tag{4}$$

$$S_h = i^2 R\text{, where} \tag{5}$$

R is the sum of electrical resistance for ohmic loss and equivalent resistance for the activation overpotential.

Numerical Method and Computational Domain

The stack tool in the present study is built by assembling a group of unit cells. Figure 2(a) shows the computational domain of a unit cell. In a unit cell the governing equations (1)-(5) are solved. The catalyst layers and the membrane are treated as solid, within which electrical and thermal conduction is considered. The GDLs as well as the fluid channels are treated as porous media. The stack tool is constructed by combining unit cells. A flow network model is implemented on the stack domain and the results appear as the inlet flow rate for the fluid channels, cf. Fig. 2(b). The stack simulation passes the flow resistance of the fluid channels back to the flow network model and this forms an iterative solution procedure. During the iteration, a current density profile is determined from potential solution for each unit cell. The current density profile is then used in evaluation of the source terms, i.e., equations (2)-(5). There are several sub-models for selected regions in the unit cell model, which are explained as follows.

Sub-models: Membrane

simplified model for the membrane is used in the present study. The model considers water transport due to electro-osmotic drag and diffusion. For simplicity the diffusivity of water in the membrane is assumed to be constant, which then makes an analytical solution of the water transport across the membrane possible:

$$\lambda = \lambda_a + \frac{e^{c_3 x}}{e^{c_3 l_m} - 1}(\lambda_c - \lambda_a) \tag{6}$$

where

$$c_1 = \frac{e^{c_3 l_m}\lambda_a - \lambda_c}{e^{c_3 l_m} - 1}, \tag{7}$$

$$c_2 = \frac{\lambda_c - \lambda_a}{e^{c_3 l_m} - 1}, \tag{8}$$

$$c_3 = \frac{2.5i}{22FD}. \tag{9}$$

Sub-model: Catalyst layer

The overpotential in the catalyst layer is expressed as a voltage drop across a resistor. The equivalent resistance at a given current density based on the Butler –Volmer equation can be computed as:

$$R_\eta = \frac{\eta}{i^N} = \frac{RT}{\alpha_c F \cdot i^N}\log\left(\frac{i^N}{a i_0 (C_{O_2})^\gamma}\right). \tag{10}$$

$i^N \equiv i$ at previous iteration

Results and Discussion

Figure 3 shows a comparison of predicted polarization for the baseline case. The computational result for a 3D model are used as the "true solution" with which all other results obtained using reduced dimensional models and simplified model are compared against. Figure 4 shows a typical result of predicted relative humidity (RH) for a 10-cell stack operated at stoichiometric factor of 5 for the cathode flow. The cathode flow and the coolant flow are from left to right in the figure, whereas the anode flow goes in the opposite direction against the cathode and coolant flows.

Conclusions

In this paper we report the numerical implementation of a stack simulation tool based on CFD framework and present some computational results obtained using the stack simulation tool. The simulation tool takes into account flow sharing, thermal and electrical coupling in a stack. The unit cell model in the simulation tool has advantages in detailed account for the flows in gas and coolant channels. The flow network model incorporated in the tool adds capabilities to predict the flow sharing in the stack headers.

Acknowledgement

The authors acknowledge the financial and technical support from Ballard Power Systems for this work.

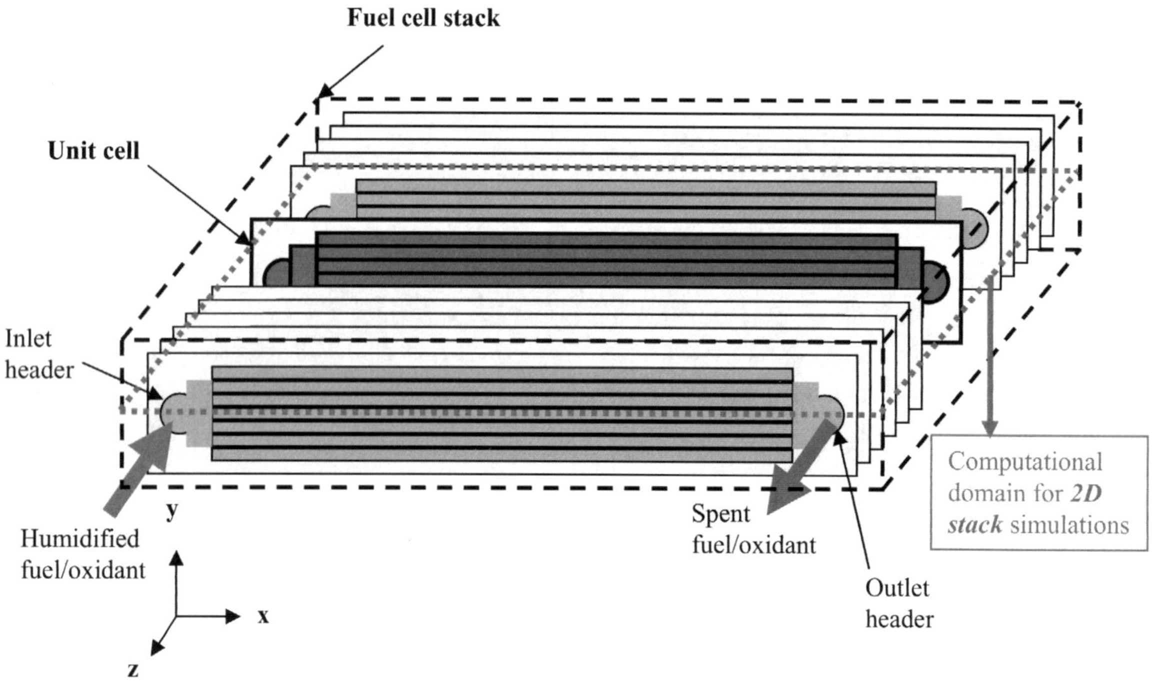

Figure 1.Schematic of a PEMFC stack

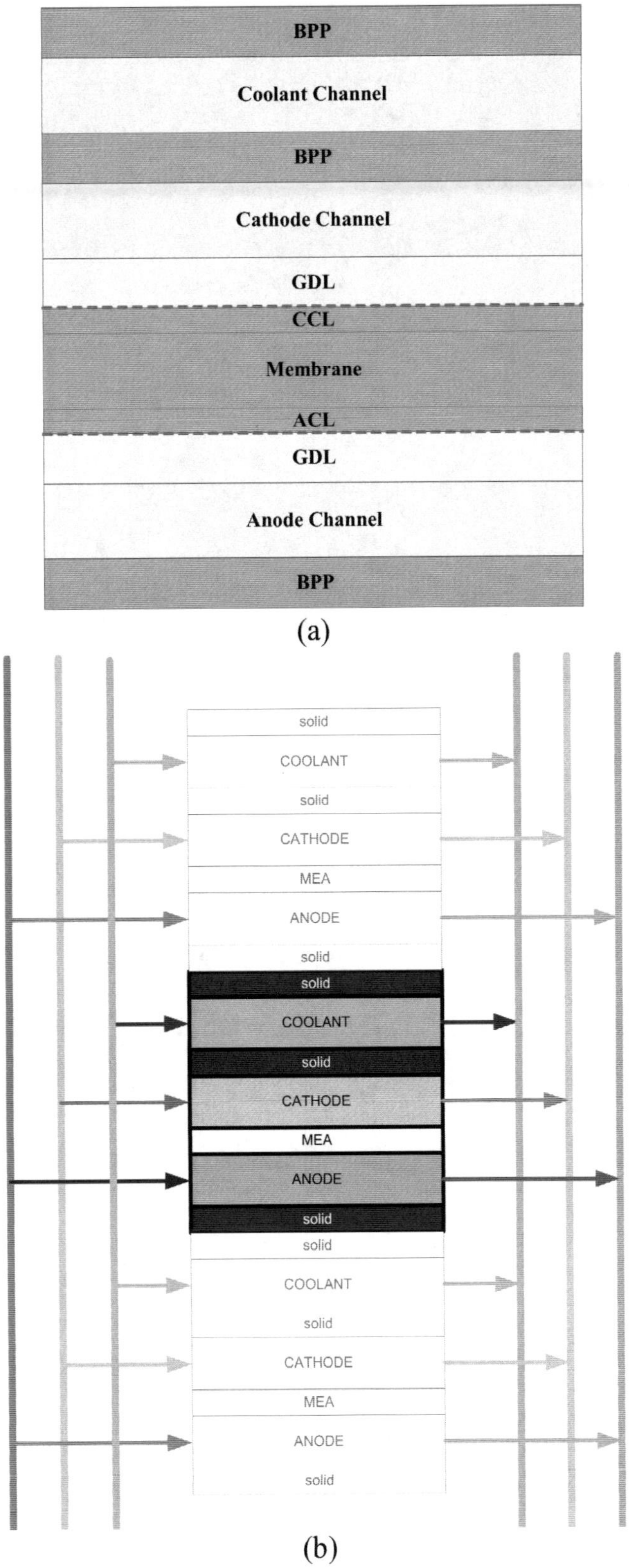

Figure 2. Schematic of the stack model (a) components of a unit cell unit (b) cell in a stack model and relation to flow network model.

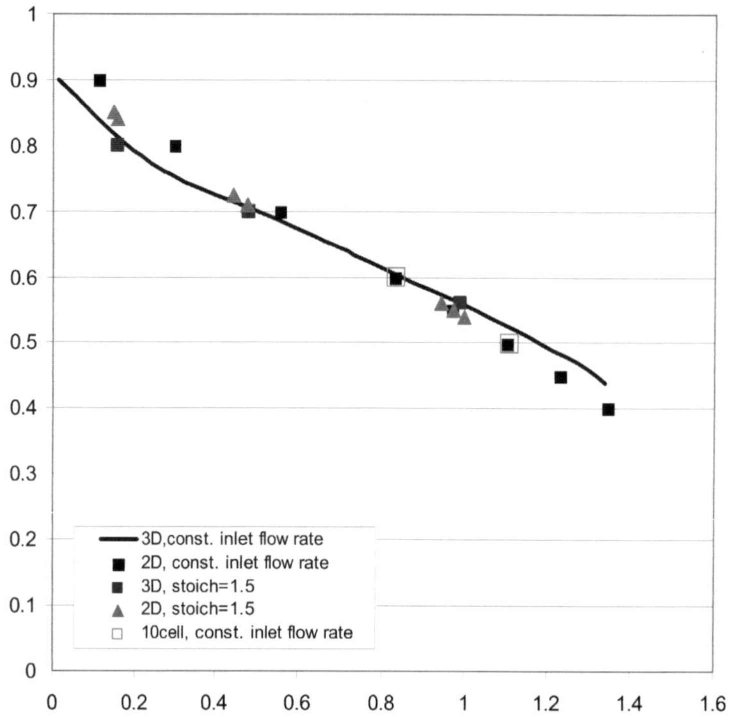

(b)

Figure 3. Comparison of predicted results, (a) species flow rate along the gas channel, (b) current density and membrane resistance

Figure 4. Relative humidity of a 10-cell stack operated at stoichiometric factor 5.

References

[1] Karimi, G., .Baschuk, J.J., Li, X. Performance analysis and optimization of PEM fuel cell stacks using flow network approach. Journal of Power Sources, 147(1-2):162–177, 2005.

[2] Kim, G.S., St-Pierre, J., Promislow, K., Wetton, B. Electrical coupling in proton exchange membrane fuel cell stacks. Journal of Power Sources, 152(1):210–217, 2005.

[3] Mackie, R., Sui, P.C. Djilali, N. Modelling and simulation of flow distribution in a PEMFC stack. ASME Fourth International Conference on Fuel Cell Science, Engineering and Technology, Irvine, CA, 2006.

[4] Springer T.E., Zawodzinski T.A., Gottesfeld S. Polymer electrolyte fuel cell model. J. Electrochem. Soc. 1991; 138(8), 2334–2342

[5] CFD-ACE+ User's Manual, v. 2006, ESI-CFD, Huntsville, AL, U.S.A.

[6] Sui P.C., Kumar S., Djilali N. Advanced computational tools for PEM fuel cell design – Part 1: Development and Base Case Simulations. J. Power Sources, in press.

[7] Sui P.C., Kumar S., Djilali N. Advanced computational tools for PEM fuel cell design Part 2. Detailed experimental validation and parametric study. J. Power Source, in press.

Water management is one of the main challenges in PEM Fuel Cells. While water is essential for membrane electrical conductivity, excess liquid water leads to flooding of catalyst layers. Despite the fact that accurate prediction of two-phase transport is key for optimal water management, understanding of the two-phase transport in fuel cells is relatively poor. Wang et. al. [1], [2] have studied the two-phase transport in the channel and diffusion layer separately using a multiphase mixture model. The model fails to accurately predict saturation values for high humidity inlet streams. Nguyen et. al. [3] developed a two-dimensional, two-phase, isothermal, isobaric, steady state model of the catalyst and gas diffusion layers. The model neglects any liquid in the channel. Djilali et. al. [4] developed a three-dimensional two-phase multicomponent model. The model is an improvement over previous models, but neglects drag between the liquid and the gas phases in the channel.

In this work, we present a comprehensive two-fluid model relevant to fuel cells. Models for two-phase transport through Channel, Gas Diffusion Layer (GDL) and Channel-GDL interface, are discussed. In the channel, the gas and liquid pressures are assumed to be same. The surface tension effects in the channel are incorporated using the continuum surface force (CSF) model. The force at the surface is expressed as a volumetric body force and added as a source to the momentum equation. In the GDL, the gas and liquid are assumed to be at different pressures. The difference in the pressures (capillary pressure) is calculated using an empirical correlations. At the Channel-GDL interface, the wall adhesion affects need to be taken into account.

SIMPLE-type methods recast the continuity equation into a pressure-correction equation, the solution of which then provides corrections for velocities and pressures. However, in the two-fluid model, the presence of two phasic continuity equations gives more freedom and more complications. A general approach would be to form a mixture continuity equation by linearly combining the phasic continuity equations using appropriate weighting factors. Analogous to mixture equation for pressure correction, a difference equation is used for the volume/phase fraction by taking the difference between the phasic continuity equations. The relative advantages of the above mentioned algorithmic variants for computing pressure correction and volume fractions are discussed and quantitatively assessed.

Preliminary model validation is done for each component of the fuel cell. The two-phase transport in the channel is validated using empirical correlations. Transport in the GDL is validated against results obtained from LBM/VOF simulation techniques [5-8]. The Channel-GDL interface transport will be validated against experiment and empirical correlation of droplet detachment at the interface.

References

[1] Y. Wang S. Basu and C.Y. Wang, Modeling two-phase flow in PEM fuel cell channels, J. Power Sources, 179:603-617, 2008.

[2] P. K. Sinha and C. Y. Wang, Liquid water transport in a mixed-wet gas diffusion layer of a polymer electrolyte fuel cell, Chem. Eng. Sci., 63:1081-1091, 2008.

[3] Guangyu Lin and Trung Van Nguyen, A two-dimensional two-phase model of a PEM fuel cell, J. Electrochem. Soc., 153(2):A372-A382, 2006.

[4] T. Berning and N. Djilali. A 3d, multiphase, multicomponent model of the cathode and anode of a PEM fuel cell, J. Electrochem. Soc., 150(12):A1589-A1598, 2003.

[5] S. Mukherjee, J.V. Cole, K. Jain, and A. Gidwani, Lattice-Boltzmann Simulations of Multiphase Flows in Gas-Diffusion-Layer (GDL) of a PEM Fuel Cell. 2008 AIChE Annual Meeting, Philadelphia, PA, Nov. 16-21 (2008).

[6] S. Mukherjee, J.V. Cole, K. Jain, and A. Gidwani, Lattice-Boltzmann Simulations of Multiphase Flows in PEM Fuel Cell GDLs and Micro-channels, 214th Meeting of Electrochemical Society, Honolulu, HI, Oct. 12-17 (2008).

[7] S. Mukherjee and J. Abraham, Investigations of Drop Impact on Dry Walls with a Lattice-Boltzmann Model, Journal of Colloid and Interface Science, 312:341-354 (2007).

[8] S. Mukherjee and J. Abraham, Liquid Crown behavior in Drop Impact on Wet Walls, Physics of Fluids, 19:052103 (2007).

An Oxide Ion and Proton Co-Ion Conducting $Sn_{0.9}In_{0.1}P_2O_7$ Electrolyte for Intermediate-Temperature Fuel Cells

Xilin Chen, Chunsheng Wang*,
Department of Chemical & Biomolecular Engineering, University of Maryland, College Park, MD, USA
Deryn Chu
U.S. Army Research Laboratory, Sensors and Electronic Devises Directorate, Adelphi, MD, USA
*Email: cswang@umd.edu
Partial work was done at the Tennessee Technological University, Cookeville, TN, USA

Abstract

The ionic conductivity of $Sn_{0.9}In_{0.1}P_2O_7$ ceramic was investigated under various atmospheres within the temperature range of 130-230°C. Similar to mixed-conductive perovskite oxides at high temperatures (such as $SrCe_{0.95}Yb_{0.05}O_{3-\alpha}$, $La_{0.9}Sr_{0.1}Ga_{0.8}Mg_{0.2}O_{3-\alpha}$ at 600-1000°C), $Sn_{0.9}In_{0.1}P_2O_7$ can conduct both protons and oxide-ions at low temperatures (130-230°C). The conductivity of $Sn_{0.9}In_{0.1}P_2O_7$ reaches 0.019 S/cm at 200°C in wet nitrogen. Its transport numbers determined by steam concentration cells are around 0.76 for proton and 0.12 for oxide-ion. The performance of direct methanol fuel cells at 170°C using mixed-ion conductive $Sn_{0.9}In_{0.1}P_2O_7$ electrolyte is higher than that at 235°C using pure proton conductive CsH_2PO_4 electrolyte. This is attributed to direct oxidation of CO at the anode by the oxide-ions generated at the cathode and moved through the $Sn_{0.9}In_{0.1}P_2O_7$ electrolyte.

Introduction

Recently, intermediate temperature solid state fuel cells, which are operated in the temperature range of 150-300°C, attracted widespread interest because of their higher CO tolerance, faster electrochemical reaction kinetics and simpler thermal and water management as compared to proton exchange membrane fuel cells (PEMFCs). Since the working temperatures of intermediate temperature fuel cells (ITFCs) are lower than that of solid oxide fuel cells, ITFCs would allow the use of oxidation-resistant metallic alloys or even plastics as interconnect materials. It would also reduce the operating cost, increase the durability, extend the service life, and permit more frequent temperature and performance cycling [1-2].

Many anhydrous proton conductors have been investigated as the electrolytes for intermediate temperature H_2/O_2 fuel cells, such as $CsHSO_4$ [3], CsH_2PO_4 [4-7], NH_4PO_3-based composites [8-13], Indium doped SnP_2O_7 [14-16]. Although the CO tolerance of CsH_2PO_4 fuel cells operating above 230°C, the transition temperature from a low- to a high-conductive phase, is much better than PEMFCs operating at 80°C, it is still lower than SOFCs because the hydrocarbon fuels can be directly electro-oxidized by the oxide ions in SOFCs [17, 18] and the operation temperatures (above 600°C) of SOFC are much higher than that of ITFCs. A method to enhance the CO tolerance of ITFCs is to use a proton and oxide ion co-ion conductive electrolyte membrane to replace the currently used proton conductive membrane. In such a cell, oxide ions are generated by the electrochemical reduction of dioxygen at the cathode and diffuse through the mixed oxide ion/proton membrane to the anode, where they may efficiently and directly oxidize any CO at the anode.

Most hydrocarbon oxidation reactions follow the Mars - van Krevelen mechanism [19-22] in which surface lattice O^{2-} is the actual oxidation agent. During these reactions, the O^{2-} reactant plays two important roles, i) being responsible for removal of H from C-H bonds to give water as in oxidative dehydrogenation, and ii) being capable of insertion into the intermediate complex to give the oxygenated product and water (mild oxidation). This mechanism is more efficient than that in the DMFC, in which methanol is directly electro-oxidized at low temperature by adsorbed hydroxide on a Pt-Ru electro-catalyst. It is the reason why an oxide-ion-conductive ceramic-supported Pt or Pt-Ru catalyst has a very high CO oxidation ability, even at low temperatures provided an oxygen (or oxide ion) source is present. It has been reported that $Sn_{0.9}In_{0.1}P_2O_7$ ITFCs can tolerate up to 10% CO in H_2; i.e., 10 %CO added to H_2 at anode did not decrease the performance of $Sn_{0.9}In_{0.1}P_2O_7$ ITFC at 250°C [14, 15]. The excellent CO tolerance at the anode for $Sn_{0.9}In_{0.1}P_2O_7$ ITFCs may attribute to the oxide ions transferring from the cathode through oxygen vacancies in $Sn_{0.9}In_{0.1}P_2O_7$ because of the existence of oxygen vacancies in $Sn_{0.9}In_{0.1}P_2O_7$ [14, 16].

In this work, the proton and oxide ion co-ion conductivity of $Sn_{0.9}In_{0.1}P_2O_7$ and the performance of $Sn_{0.9}In_{0.1}P_2O_7$ fuel cells with methanol and/or hydrogen fuel were investigated. A fuel cell using a co-ion conducting $Sn_{0.9}In_{0.1}P_2O_7$ membrane, operating at temperatures of 170-200°C, combined the advantages of both the SOFC and PEMFC, and complemented existing proton and oxide-ion fuel cell systems. The goal of introducing oxide ion transport in a proton conducting ceramic is to oxidize CO and methanol in an ITFC, which is different from prior works on mixed proton/oxide ion conducting perovskite oxides (such as $SrCe_{0.95}Yb_{0.05}O_{3-\alpha}$ [23], $La_{0.9}Sr_{0.1}Ga_{0.8}Mg_{0.2}O_{3-\alpha}$ [24]), for which proton conductivity is thermally activated at a lower temperature than that required for oxide ion conduction [25] and hence enables a lower SOFC operating temperature.

Experimental

In^{3+}-doped SnP_2O_7 was prepared using SnO_2 (Sigma-Aldrich 99.9%), In_2O_3 (Aldrich 99.99%) and H_3PO_4 (Fisher Scientific 85%) to form $Sn_{0.9}In_{0.1}P_2O_7$. The starting materials were mixed and held with stirring at 300°C until the mixture formed a paste with a high viscosity. The pastes were calcined at 650°C for 2.5 hours and then ground into powders with a mortar and pestle. The compound was pressed at 250MPa into pellets and then coated with Ag for conductivity measurement and with Pt catalyst for fuel cell testing. The pellets were 13 mm in diameter and around 0.8 mm in thickness. For comparison, CsH_2PO_4 was also prepared as described in [4] using Cs_2CO_3 (Sigma-Aldrich 99%) and H_3PO_4 (Fisher Scientific 85%) as precursors. The crystalline structure of the compounds was analyzed by X-ray diffraction (PANalytical X'Pert Pro MPD diffractometer with Cu radiation) and the micro-morphology of the powders and pellets was observed by scanning electron microscopy (SEM) images using a FEI Quanta 200 with Electron Diffraction Spectrometer (EDS).

Both the anode and cathode contained two overlapped conductive porous layers: a diffusion layer and a catalyst layer. The diffusion layer was formed from carbon cloth (E-TEK, B1A) which was smoothed by coating a mixture of 85 wt.% carbon powder (Vulcan XC72) and 15 wt.% polytetrafluoro-ethylene (PTFE, Aldrich, 60% dispersion). The carbon powder loading was 2 mg/cm^2. The catalyst layer was formed from either $Sn_{0.9}In_{0.1}P_2O_7$ or CsH_2PO_4 (35wt.%), 40wt.% Pt/C catalyst (50wt.%, E-TEK) and PTFE (15 wt.%). The catalyst ink was prepared by mixing $Sn_{0.9}In_{0.1}P_2O_7$ (or CsH_2PO_4), distilled water, Pt/C catalyst, isopropanol (Fisher Scientific, 99.9%) and PTFE. The ink was brushed onto the diffusion layer and dried to form an electrode (anode or cathode). The platinum loading was 4 mg/cm^2. The electrolyte membrane ($Sn_{0.9}In_{0.1}P_2O_7$ or CsH_2PO_4) was sandwiched by the anode and cathode to form a single cell. Humidification of gas was achieved by saturating nitrogen/hydrogen/oxygen with water at a given temperature, resulting in a given water partial pressure, such as 76°C for 40% H_2O. In order to add methanol into fuel gases, hydrogen/nitrogen was saturated with a mixture of methanol and water at a given temperature.

Conductivity measurements were carried out using a Salartron (SI1287 and SI1260) electrochemical impedance system in the frequency range of 0.001-10^6 Hz. At each temperature, the sample was held about 30 minutes prior to measurement.

Results and discussions

Figure 1 shows typical XRD patterns for $Sn_{0.9}In_{0.1}P_2O_7$ and CsH_2PO_4 at room temperature. The peaks observed are identical to those reported in PDF# 00-29-1352 (SnP_2O_7) and PDF# 00-35-0746

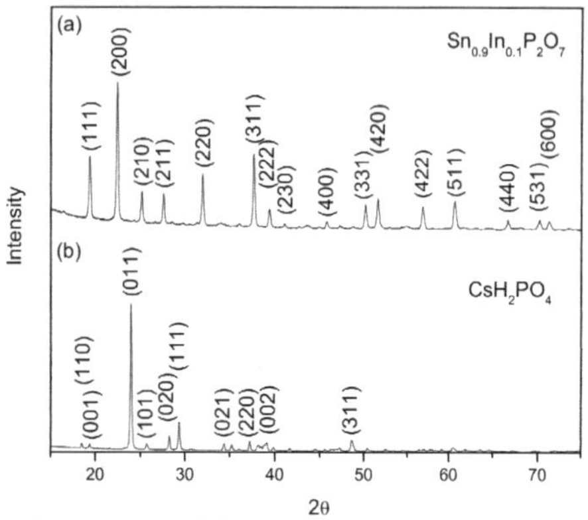

Figure 1 X-ray diffraction patterns for (a) $Sn_{0.9}In_{0.1}P_2O_7$ and (b) CsH_2PO_4

Figure 2 Micro-morphologies of (a) $Sn_{0.9}In_{0.1}P_2O_7$ powder, (b) CsH_2PO_4 powder, (c) cross-section for $Sn_{0.9}In_{0.1}P_2O_7$ pellet, and (d) cross-section for CsH_2PO_4 pellet.

(CsH_2PO_4). SnP_2O_7 doped with 10% In^{3+} showed the same structure as the undoped SnP_2O_7, characterized by a cubic or pseudo-cubic structure with SnO_6 octahedral at the corners and P_2O_7 units at the edges [26]. However, M. Nagao et al. reported that 10% In^{3+} doping increased the lattice constant from 7.945 Å to 7.950 Å [16] and created oxygen vacancy defects [14, 16]. And the chemical composition of $Sn_{0.9}In_{0.1}P_2O_7$ was confirmed by Energy Dispersive Analysis of X-rays (EDAX). All peaks in Figure 1b could be indexed to CsH_2PO_4, indicating the synthesized material is pure CsH_2PO_4.

Figure 2 shows the morphologies of powders and morphologies of cross-sections for $Sn_{0.9}In_{0.1}P_2O_7$ and CsH_2PO_4 pellets. The cross-sections were obtained by cutting the pellets with a sharp blade. Although the particle size of $Sn_{0.9}In_{0.1}P_2O_7$ is smaller than that of CsH_2PO_4, (Figures 2 a and b), the porosity of the former is higher than the latter, i.e., the density of the CsH_2PO_4 pellet is higher than that of $Sn_{0.9}In_{0.1}P_2O_7$ (Figure 2 c and d).

Figure 3 shows the conductivities of $Sn_{0.9}In_{0.1}P_2O_7$ and CsH_2PO_4 at different temperatures in humidified nitrogen. The conductivity of $Sn_{0.9}In_{0.1}P_2O_7$ continuously increased with increasing water vapor pressure from 3.2 to 40% (Fig. 3a). The conduction behavior of $Sn_{0.9}In_{0.1}P_2O_7$ at different temperatures is different from superprotonic character of CsH_2PO_4 that shows a sharp increase in the conductivity of a few orders of magnitude by a structural transition from a low- to a high-temperature phase (Fig. 3). The maximum protonic conductivities of $Sn_{0.9}In_{0.1}P_2O_7$, achieved at around 200°C in 40% H_2O, is 0.019 S/cm, which is similar to that (0.02 S/cm) of CsH_2PO_4 at 240°C. The thermal variations of proton conductivities for $Sn_{0.9}In_{0.1}P_2O_7$ and CsH_2PO_4 were consistent with reported results [4, 14], although the conductivity of $Sn_{0.9}In_{0.1}P_2O_7$ is slightly lower than reported values [14]. The reason why the conductivity decreases after 200°C is not clear. It cannot be attributed to the escape of water produced by the reversal of reaction 1 because the evolution of water vapor from $Sn_{0.9}In_{0.1}P_2O_7$ starts above 260°C [16]. If it is due to the escape of water, the peak position should not be fixed at 200°C but move to a higher temperature with higher water partial pressure.

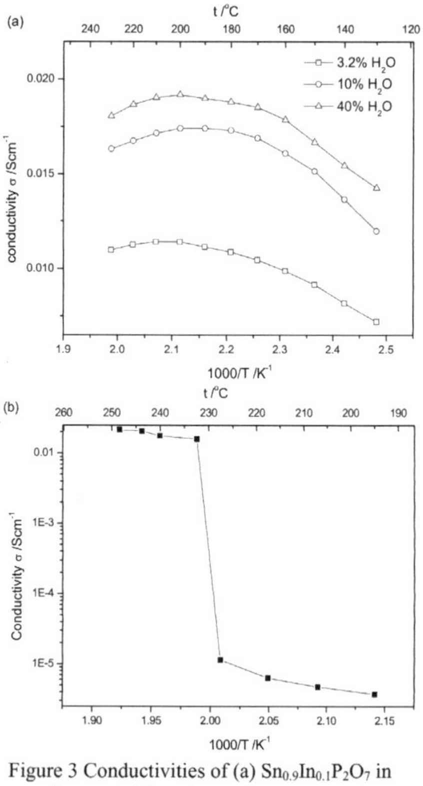

Figure 3 Conductivities of (a) $Sn_{0.9}In_{0.1}P_2O_7$ in wet nitrogen with various water vapor pressures, and (b) CsH_2PO_4 in 40% H_2O/N_2.

The interaction (reaction 1) between water vapor and oxygen vacancies (and/or electron holes) may be responsible for protonic conduction in $Sn_{0.9}In_{0.1}P_2O_7$ [14, 16], which is similar to the mechanism of proton incorporation in perovskite oxides such as $SrCe_{0.95}Yb_{0.05}O_{3-\alpha}$.. The increase in conductivity of $Sn_{0.9}In_{0.1}P_2O_7$ with humidity (Fig. 3) supports the hypothesis of the proton conducting mechanism. During the operation of fuel cells, protons generated at anode (reaction 2) move through the $Sn_{0.9}In_{0.1}P_2O_7$ membrane to cathode to react with oxygen (reaction 3), whereas oxide ions transits in the opposite direction from cathode (reaction 4) to anode to oxidize the CO or CH_3OH (reaction 5). CO and methanol can also be chemically oxidized with water through water-gas-shift reaction (reaction 6) and reforming (reaction 7) at intermediate temperatures.

Proton conduction in $Sn_{0.9}In_{0.1}P_2O_7$

$$H_2O(g) + V_O^{\bullet\bullet} \rightarrow 2H_i^{\bullet} + O_O^{\times}$$

$$\text{and/or } H_2O(g) + 2h^{\bullet} \rightarrow 2H_i^{\bullet} + \tfrac{1}{2}O_2 \quad (1)$$

Anode reaction in $Sn_{0.9}In_{0.1}P_2O_7$ fuel cells

$$H_2 \rightarrow 2H_i^{\bullet} + 2e^- \quad (2)$$

Cathode reaction in $Sn_{0.9}In_{0.1}P_2O_7$ fuel cells

$$\tfrac{1}{2}O_2 + 2H_i^{\bullet} + 2e^- \rightarrow H_2O \quad (3)$$

Oxide ion conduction in $Sn_{0.9}In_{0.1}P_2O_7$

$$\tfrac{1}{2}O_2 + V_O^{\bullet\bullet} + 2e^- \rightarrow O_O^{\times} \quad (4)$$

Direct CO and methanol oxidation by oxide ion in anode

$$O_O^{\times} + [R] \rightarrow V_O^{\bullet\bullet} + [P] + 2e^- \quad (5)$$

CO oxidation in anode through water-gas-shift reaction

$$H_2O + CO \rightarrow CO_2 + H_2 \quad (6)$$

CH_3OH oxidation in anode through reforming

$$CH_3OH + H_2O \rightarrow CO_2 + 3H_2 \quad (7)$$

Here, $h^{\bullet}$, $H^{\bullet}$, $V_O^{\bullet\bullet}$, $O_O^{\times}$, $[R]$, $[P]$ and e^- denote an electron hole, a proton, an oxygen vacancy, a lattice oxide ion, reductant (CO and CH_3OH), reduction product and an electron, respectively.

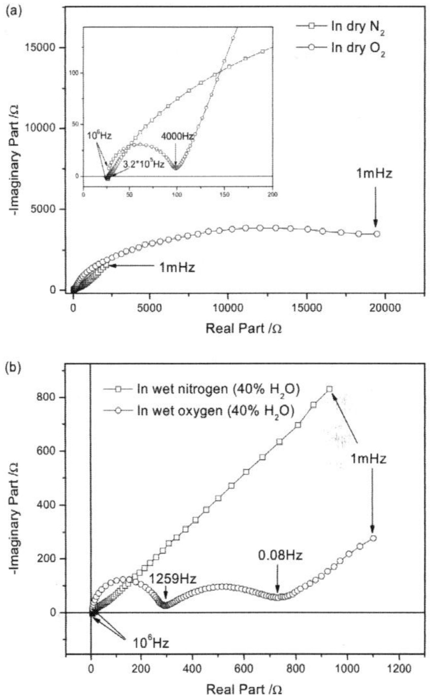

Figure 4 Impedance plots of $Sn_{0.9}In_{0.1}P_2O_7$ in (a) dry and (b) wet condition at 170°C, electrolyte thickness: 0.79mm for dry oxygen, 0.80mm for dry nitrogen, 0.88 mm for wet oxygen, 0.80mm for wet nitrogen.

Another possibility for proton conduction in $Sn_{0.9}In_{0.1}P_2O_7$ is the residual phosphoric acid on the surface of material. If phosphoric acid exists on the surface of material, it should be dissolved into water when the $Sn_{0.9}In_{0.1}P_2O_7$ particles are soaked in the water, thus the washed $Sn_{0.9}In_{0.1}P_2O_7$ should have lower conductivity than that of unwashed $Sn_{0.9}In_{0.1}P_2O_7$. However, the conductivity of washed $Sn_{0.9}In_{0.1}P_2O_7$ shows similar conductivity with unwashed one, indicating that proton conduct in $Sn_{0.9}In_{0.1}P_2O_7$ is mainly through water- oxygen vacancy reaction. More experiments are still needed to elucidate the exact role of residual phosphoric acid in proton conduction of $Sn_{0.9}In_{0.1}P_2O_7$ if it exists.

To check the conductivity of $Sn_{0.9}In_{0.1}P_2O_7$ electrolyte in fuel cells, the impedances of $Sn_{0.9}In_{0.1}P_2O_7$ membranes were measured in O_2 and N_2 with and without the presence of water (Fig. 4). Comparing the impedances of $Sn_{0.9}In_{0.1}P_2O_7$ measured in dry N_2 and dry O_2 (Fig. 4a). The low-frequency-line did not intersect the real axis, which indicates that $Sn_{0.9}In_{0.1}P_2O_7$ is an electronic insulator. Adding water to oxygen gas results in an even lower electronic conductivity, as evidenced by almost straight line in the low frequency region (Fig. 4b). Fig. 4 clearly demonstrated that $Sn_{0.9}In_{0.1}P_2O_7$ is an electronic insulator at temperatures below 200°C whether in a dry or wet atmosphere.

The maximum protonic conductivities of 0.019 S/cm is achieved at around 200°C in 40% H_2O. To avoid the conductivity drop, the $Sn_{0.9}In_{0.1}P_2O_7$ membranes and $Sn_{0.9}In_{0.1}P_2O_7$ fuel cells were tested at 170°C (i.e., below 200°C). Figure 5 shows the time dependence of conductivity of $Sn_{0.9}In_{0.1}P_2O_7$ at 170°C and CsH_2PO_4 at 235°C in wet nitrogen with 40% water vapor. It is clear that the conductivities of both $Sn_{0.9}In_{0.1}P_2O_7$ and CsH_2PO_4 are very stable. Therefore, both $Sn_{0.9}In_{0.1}P_2O_7$ and CsH_2PO_4 are suitable electrolytes for intermediate temperature fuel cells.

Based on reactions 1 and 4, $Sn_{0.9}In_{0.1}P_2O_7$ should be a mixed proton and oxide ion conductor in humidified gas. In order to confirm $Sn_{0.9}In_{0.1}P_2O_7$ is a mixed-conducting conductor under humidified conditions, the transport numbers of protons and oxide ions were determined by measuring the electromotive force, EMF, of steam concentration cells H_2/O_2 (P_{H_2O}=0.6KPa), Pt/C|$Sn_{0.9}In_{0.1}P_2O_7$|Pt/C, H_2/O_2 (P_{H_2O}=3.2KPa). The Nernst potential of the steam concentration cell can be determined by the following equations [28]:

$$E_0 = (RT/4F)[t_{ion} \ln(P^{II}_{O_2}/P^{I}_{O_2}) - 2t_{H^+} \ln(P^{II}_{H_2O}/P^{I}_{H_2O})] \quad (8)$$

where $t_{ion} = t_{H^+} + t_{O^{2-}}$ is the ionic transport number and the carrying gas is oxygen. Or

$$E_0 = (RT/2F)[-t_{ion} \ln(P^{II}_{H_2}/P^{I}_{H_2}) + t_{O^{2-}} \ln(P^{II}_{H_2O}/P^{I}_{H_2O})] \quad (9)$$

where the carrying gas is hydrogen.

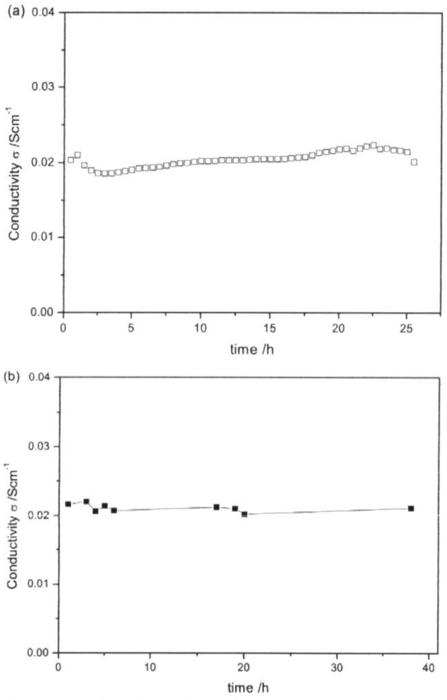

Figure 5 (a) Time dependence of conductivities of $Sn_{0.9}In_{0.1}P_2O_7$ at 170°C 40% H_2O/N_2 and (b) that of CsH_2PO_4 at 235°C in 40% H_2O/N_2

Since the pressures of the carrying gases (oxygen) at the both sides of the steam concentration cell are very close and transport number for protons is larger than that for oxide ions (i.e. $2t_{H^+} >> t_{H^+} + t_{O^{2-}}$), the first parts of equations (8) are negligible. Therefore, the equations (8) can be expressed as: $E_0 = -t_{H^+}(RT/2F)\ln(P^{II}_{H_2O}/P^{I}_{H_2O})$, and the proton transport number, t_{H^+}, can be calculated from the measured E_0, $t_{H^+} = E_0/E_{th}$ [29]. Here $E_{th} = (RT/2F)\ln(P^{II}_{H_2O}/P^{I}_{H_2O})$ is defined as the theoretical EMF. As shown in Figure 6a, the proton transport numbers are 0.7-0.8 (average 0.76), depending on the temperatures. After switching the carrying gas from oxygen to hydrogen, the oxygen-ion transport number $t_{O^{2-}}$ at different temperatures can be calculated from equation (9) using the measured E_0 values and calculated proton transport numbers at the same temperatures. As shown in Figure 6b, the oxygen-ion transport numbers are around 0.12. The total ionic transport number (0.88) of $Sn_{0.9}In_{0.1}P_2O_7$ is less than unity, which is probably attributed to gas leakage in the steam concentration cells due to the low density of the membrane (Figure 2). The gas leakage will result in a lower measured EMF than the theoretical value, and less than unity for the total ionic transport numbers. The conductivity of oxide ions in $Sn_{0.9}In_{0.1}P_2O_7$ at 200°C is around 10^{-3} S/cm, which is close to the oxide ion conductivities of $La_{0.8}Sr_{0.2}Ga_{0.8}Mg_{0.2}$ (LSGM) [30] and $Bi_4Ti_{0.2}V_{1.8}O_{10.7}$ [31] at 250°C.

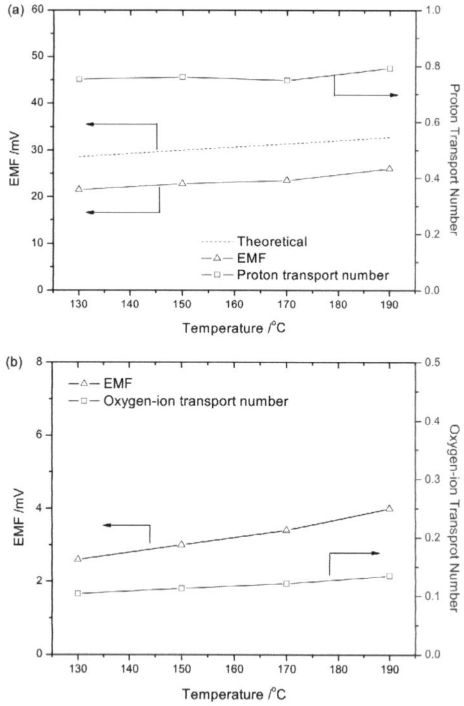

Figure 6 EMF values of $Sn_{0.9}In_{0.1}P_2O_7$ membrane galvanic cells at different atmospheres. (a) Anode gas=O_2 (P_{H2O}=3.2KPa); Cathode gas=O_2 (P_{H2O} =0.6KPa). (b) Anode gas=H_2 (P_{H2O} =0.6KPa); Cathode gas=H_2 (P_{H2O} =3.2KPa); electrolyte thickness: 1.4mm.

The carbon monoxide generated by reforming reaction would dilute the hydrogen fuel and/or poison the catalyst in anode even at intermediate temperature and make fuel cell performance worse. Since $Sn_{0.9}In_{0.1}P_2O_7$ is a proton and oxide ion co-ion conductor and the oxide ion will oxidize the CO (reaction 5 can be particularized as $O_O^{\times} + CO \rightarrow V_O^{\bullet\bullet} + CO_2 + 2e^-$), the exhaust CO can become as a fuel in the direct methanol $Sn_{0.9}In_{0.1}P_2O_7$ fuel cells, and thus the direct methanol $Sn_{0.9}In_{0.1}P_2O_7$ fuel cells should perform better than direct methanol CsH_2PO_4 fuel cells.

To test the capability of direct oxidation of CO by oxide ions, the performance of a Pt/C fuel cell using a mixed proton/oxide ion conducting $Sn_{0.9}In_{0.1}P_2O_7$ electrolyte was investigated in the temperature range 130-170°C and compared with a fuel cell using a pure proton conductor CsH_2PO_4 electrolyte at 235°C. Cell voltages and power densities of two cells are shown in Figures 7 and 8 as a function of current density. As expected, the power density of the $Sn_{0.9}In_{0.1}P_2O_7$ fuel cell increased with temperature and the maximum power densities at 170°C were 14.7mW/cm^2 for hydrogen, 14.9mW/cm^2 for a mixture of hydrogen and methanol and 4.0mW/cm^2 for methanol. Due to the fast reaction kinetics at a comparatively high operating temperature (235°C) for the CsH_2PO_4 fuel cell, the maximum power density (17.3mW/cm^2) of the H_2/O_2 CsH_2PO_4 fuel cell was slightly higher than the H_2/O_2 $Sn_{0.9}In_{0.1}P_2O_7$ fuel cell, even though the ionic conductivities of the two electrolytes are very close. However, the power densities of the CsH_2PO_4 fuel cell at 235°C using hydrogen-methanol (13.7mW/cm^2) and methanol (2.1mW/cm^2) were much lower than that of the $Sn_{0.9}In_{0.1}P_2O_7$ fuel cell at 170°C. When 15% methanol was added into hydrogen, the power density of the CsH_2PO_4 fuel cell lost 21% power density, decreasing from 17.3mW/cm^2 to 13.7mW/cm^2, but the power density of the $Sn_{0.9}In_{0.1}P_2O_7$ fuel cell remained the same. When the fuel was switched from the H_2-CH_3OH mixture to pure methanol, the power density for the CsH_2PO_4 fuel cell had an 85% drop, from 13.7mW/cm^2 to 2.1mW/cm^2, while the drop for the $Sn_{0.9}In_{0.1}P_2O_7$ fuel cell was only 75%, from 14.9mW/cm^2 to 4.0mW/cm^2. Since both the amount of platinum loading in two fuel cell cells and ionic conductivity of two electrolytes (CsH_2PO_4 and $Sn_{0.9}In_{0.1}P_2O_7$) were the same, even the CsH_2PO_4 cell has faster electrochemical reaction kinetics than that of the $Sn_{0.9}In_{0.1}P_2O_7$ cell because of its higher operation temperature, the difference in performance must come from the electrolytes. The higher performance of methanol fuel cells using $Sn_{0.9}In_{0.1}P_2O_7$ as an electrolyte at 170°C compared to using a CsH_2PO_4 electrolyte at 235°C strongly indicated that CO was effectively oxidized by the oxide ions, resulting in enhancement of the fuel cell performance.

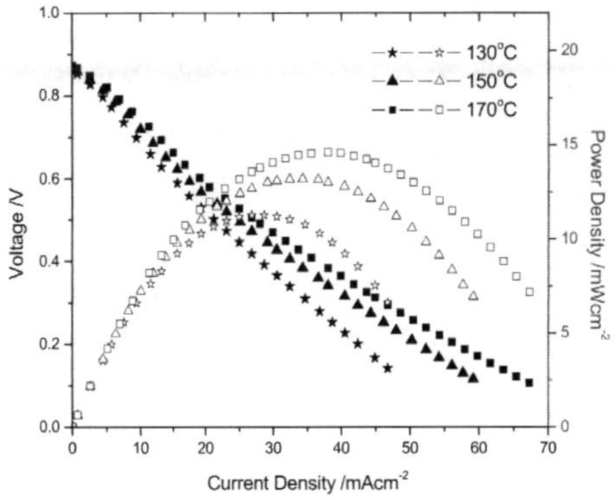

Figure 7 Fuel cell performance at different temperature using $Sn_{0.9}In_{0.1}P_2O_7$ as electrolyte and hydrogen as fuel, electrolyte thickness: 0.78mm

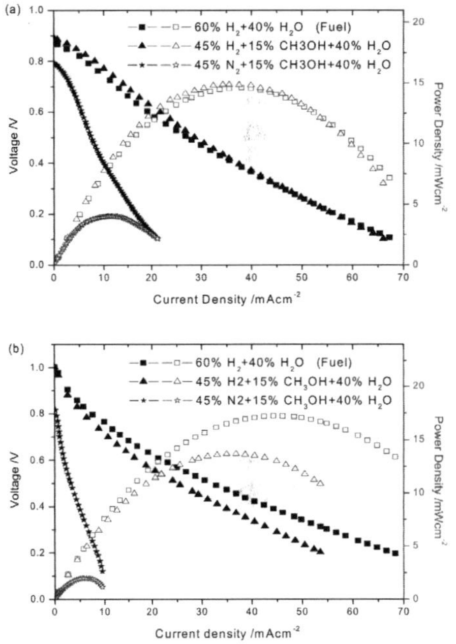

Figure 8 Fuel cell performance with various fuels, electrolytes: (a) $Sn_{0.9}In_{0.1}P_2O_7$, thickness: 0.78mm, 170°C (b) CsH_2PO_4, thickness: 0.89mm, 235°C

Although the CsH_2PO_4 fuel cell has a lower power density than the $Sn_{0.9}In_{0.1}P_2O_7$ fuel cell when methanol is present in the fuel, its open circuit voltage is higher. This is due to the higher density of the CsH_2PO_4 membrane as shown in Figure 2, and the higher density may be another reason that the power density of CsH_2PO_4 is higher than $Sn_{0.9}In_{0.1}P_2O_7$ in a H_2/O_2 fuel cell, even though they have very similar conductivities.

Conclusions

A proton and oxide ion co-ion conductor, $Sn_{0.9}In_{0.1}P_2O_7$, was synthesized and characterized. The conductivity of $Sn_{0.9}In_{0.1}P_2O_7$ and performance of $Sn_{0.9}In_{0.1}P_2O_7$ membrane fuel cell were investigated under various atmospheres in the temperature range of 130-230°C and compared with pure proton conductive CsH_2PO_4 and CsH_2PO_4 membrane fuel cells. The conductivity of $Sn_{0.9}In_{0.1}P_2O_7$ reached 0.019 S/cm at 200°C in wet nitrogen, and the transport numbers determined by steam concentration cells were 0.76 for protons and 0.12 for oxide ions. The higher performance of a methanol $Sn_{0.9}In_{0.1}P_2O_7$ fuel cell at 170°C compared to a methanol CsH_2PO_4 fuel cell at 235°C was attributed to the direct oxidation of CO at the anode by the oxide ions transported through the $Sn_{0.9}In_{0.1}P_2O_7$ electrolyte. Due to the unique concept of oxidation of CO by oxide ion transported from the cathode, and preliminary results on a co-ion (proton/oxide ion) conductive fuel cell with methanol fuel, the co-ion conducting membrane fuel cell, operating at around 200°C, may be the next-generation methanol fuel cell and reforming fuel cell because it combines the advantages of both SOFC and PEMFC.

Acknowledgement

This material is based upon work supported by the U. S. Army Research Laboratory and the U. S. Army Research Office under grant number W911NF-06-1-0187. The XRD analysis is sponsored by the Assistant Secretary for Energy Efficiency and Renewable Energy, Office of FreedomCAR and Vehicle Technologies, as part of the High Temperature Materials Laboratory User Program, Oak Ridge National Laboratory, managed by UT-Battelle, LLC, for the U.S. Department of Energy under contract number DE-AC05-00OR22725

References

1. T. Norby, Solid State Ionics, 125 (1999) 1-11
2. T. Norby, Nature, 410 (2001) 877-878
3. S. M. Haile, D. A. Boysen, C. R. I. Chisholm, R. B. Merle, Nature, 410 (2001) 910-913
4. S. M. Haile, C. R. I. Chisholm, K. Sasaki, D. A. Boysen, T. Uda, Faraday Discuss, 134 (2007) 17-39
5. T. Uda, S. M. Haile, Electrochemical and Solid-State Letters, 8 (5) (2005) A245-A246
6. D. A. Boysen, T. Uda, C. R. I. Chisholm, S. M. Haile, Science, 303 (2004) 68-70
7. T. Uda, D. A. Boysen, C. R. I. Chisholm, S. M. Haile, Electrochemical and Solid-State Letters, 9(6) (2006) A261-A264
8. M. Cappadonia, O. Niemzig, U. Stimming, Solid State Ionics, 125 (1999) 333-337
9. S. Haufe, D. Prochnow, D. Schneider, O. Geier, D. Freude, U. Stiming, Solid State Ionics, 176 (2005) 955-963
10. T. Matsui, S. Takeshita, Y. Iriyama, T. Abe, Z. Ogumi, Journal of The Electrochemical Society, 152 (1) (2005) A167-A170
11. X. Chen, Z. Huang, C. Xia, Solid State Ionics, 177 (2006) 2413-2416
12. X. Chen, X. Li, S. Jiang, C. Xia, U. Stimming, Electrochimica Acta, 51 (2006) 6542-6547
13. X. Chen, C. Xia, U. Stimming, Electrochimica Acta, 52 (2007) 7835-7840
14. M. Nagao, A. Takeuchi, P. Heo, T. Hibino, M. Sano, A. Tomita, Electrochemical and Solid-State Letters, 9 (3) (2006) A105-A109
15. P. Heo, H. Shibata, M. Nagao, T. Hibino, M. Sano, Journal of The Electrochemical Society, 153 (5) (2006) A897-A901
16. M. Nagao, T. Kamiya, P. Heo, A. Tomita, T. Hibino, and M. Sano, Journal of The Electrochemical Society, 153 (2006) A1604-A1609
17. E. P.Murray, T. Tsai, S.A. Barnett, Nature, 400 (1999) 649-651
18. S. Park, J. M. Vohs, R. J. Gorte, Nature, 404 (2000) 265-267

19. P. Mars, D.W. van Krevelen, Chemcal Engineering Science, 3 (Spec. Suppl.) (1954) 41-59
20. C. Doornkamp, V. Ponec, Journal of Molecular Catalysis A: Chemical, 162(1) (2000) 19-32
21. F. Bertinchamps, E.M. Gaigneaux, Catalysis Today, 91-91 (2004) 105-110
22. H. Over, Y.D. Kim, A.P. Seitsonen, S. Wendt, E. Lundgren, M. Schmid, P. Varga, A. Morgante, G. Ertl, Science, 287 (2000) 1474-1478
23. H. Uchida, N. Maeda, H. Iwahara, Solid State Ionics, 11 (1983) 117-124
24. G. Ma, F. Zhang, J. Zhu, G. Meng, Chemistry of Materials, 18 (2006) 6006-6011
25. B. Zhu, X. Liu, T. Schober, Electrochemistry Communications, 6 (2004) 378-383
26. R. K. B. Gover, N. D. Withers, S. Allen, R. L. Withers, J. S. O. Evans, Journal of Solid State Chemistry 166 (2002) 42-48
27. J. Otomo, N. Minagawa, C. Wen, K. Eguchi, H. Takahashi, Solid State Ionics, 156 (2003) 357-369
28. J. Guan, S. E. Dorris, U. Balachandran, M. Liu, Solid State Ionics, 100 (1997) 45-52
29. D. P. Sutija, T. Norby, P. Bjrönbom, Solid State Ionics, 77 (1995) 167-174
30. J. W. Fergus, Journal of Power Sources, 162 (2006) 30-40
31. M. J. Godinho, P. R. Bueno, M. O. Orlandi, E. Longo, Materials Letters, 57 (2003) 2540-2544

648c An Efficient Approach to the Management of Liquid Water In PEM Fuel Cells

Erin Kimball, Jay B. Benziger, and Yannis G. Kevrekidis. Department of Chemical Engineering, Princeton University, Princeton, NJ 08544

The operation of polymer electrolyte membrane (PEM) fuel cells requires a sensitive balance of the amount of water inside the cell. There must be enough water so that the membrane is well hydrated and has a high proton conductivity, but not too much so that liquid water accumulates and blocks the flow of reactants to the catalyst. In practice, this is achieved by increasing the flow rates of the gases, so that any excess water is simply blown out of the cell, and then humidifying the feeds to counteract any drying of the membrane, especially near the inlets. This solution adds a need for humidifiers and a recycling system to keep the fuel efficiency high, both increasing the cost and complexity of the fuel cell system. Our work takes a different approach by using low flow rates and exploiting the material properties and fundamental physics of the system.

Previous work in our group has shown the importance of two key aspects of the operation of fuel cells high fuel utilization: the structure of the gas diffusion layer (GDL) and the effects of gravity (Kimball, et al., AIChE J. 54(5), 2008). A model fuel cell with a single transparent straight flow channel and segmented anode allowed for the direct correlation of liquid water movement with the dynamic current distribution along the channel. The stability of operation was determined largely by the degree with which gravity assisted in removing the liquid water from the cell, ranging from small periodic fluctuations to huge oscillations with periods of only seconds. These effects were then coupled with the effects of the hydrophobicity of the GDL. Water flows through only the largest pores of the GDL with the size of the droplets that emerge on the surface determined by the size of the pore. This was exploited by forcing the largest pore to be in a certain location along the channel and either under the land or under the channel. The mechanism of how the water flowed into and along the channel affected the magnitude and frequency of fluctuations in the local currents as well as the uniformity of the current distribution along the channel. The effects of gravity were amplified when the large pore was under the channel, but diminished with the large pore under the land. These results have implications for application-specific system design, an issue that is not widely discussed, and for improving the durability of the membrane and electrodes, one of the major hindrances to increasing the mainstream use of fuel cells.

Web Page: pemfc.princeton.edu/index.htm

648d Magnetic Resonance Imaging as a Diagnostic Tool for Fuel Cell Water Management

Zachary Dunbar, Chemical Engineering, University of Illinois at Urbana-Champaign, 600 S Mathews, Urbana, IL 61801 and Rich Masel, Chemical Engineering, University of Illinois, Urbana-Champaign, 600 South Mathews, Room 204, Urbana, IL 61801.

Water management remains a significant challenge for the commercialization of PEM fuel cells. Many mathematical models have been created to gain insight into the effects of water distribution on cell performance. However, model verification remains difficult due to the challenging nature of acquiring relevant experimental data about water generation and movement inside an operating fuel cell.

In this work, magnetic resonance imaging (MRI) is demonstrated as a strong diagnostic tool to gather information about water generation and transport inside an operating PEM fuel cell. The MRI technique is used to gather quantitative water distribution data in three dimensions. Acquisition times are as fast as 8 seconds, allowing observation of dynamic processes within the flow channels. Flow channel depth and geometry are varied, and the effect on water distribution and overall cell performance measured.

The MRI technique is shown to be a strong diagnostic tool for fuel cell researchers, due to its fast, quantitative and 3-dimensional spatial data acquisition properties.

648e U.S. Navy Desulfurization Adsorbent Development

John Heinzel[1], Ian Peek[2], Donald Hoffman[3], and John Kuseian[2]. (1) NSWCCD Code 985, Naval Sea Systems Command - Philadelphia, 5001 S. Broad St, Philadelphia, PA 19112, (2) Code 985, Naval Sea Systems Command - Philadelphia, 5001 S. Broad St, Philadelphia, PA 19112, (3) Code 331, Office of Naval Research, 875 N. Randolph St., Arlington, VA 22203

The U.S. Navy's Office of Naval Research (ONR) is sponsoring advanced research in a variety of component areas, with the goal of obtaining a three to four time improvement in overall system power density. The technology development programs ongoing as part of the ONR advanced fuel cell program are addressing numerous areas of process improvements. One of the most critical is the ability to remove sulfur from F-76 and JP-5 Navy logistics fuel, in liquid phase at ambient temperature and pressure, with high regeneration capability over high numbers of cycles. Such removal of refractory sulfur compounds is a challenge due to the nature of the molecules themselves, as well as their high concentrations, up to 5000 ppm of sulfur. The ability to remove sulfur from fuel at such conditions is important, as it provides benefits in process simplification and system safety. Liquid-phase desulfurization allows systems to become far more compact, due to the removal of pressure vessels, large valve and actuator assemblies and high inventory reactors, which contain large amounts of vaporized fuel and other combustibles.

The goal of the sorbent development program is to create new and improved desulfurization sorbents which have ten or more times the capacity of current sorbents. The previous state-of-the-art sorbents at the start of the program exhibited capacities of approximately 0.5 mg S/g sorbent, when utilized at ambient conditions. As various materials may be especially well suited to specific types of sulfur compounds, the U.S. Navy is performing development work, while also partnering with Auburn University, the Pennsylvania State University and the University of Michigan, to explore a wide variety of novel sorbent types.

This paper describes the most current updates and results on efforts ongoing as part of the ONR advanced sorbent work. The paper describes new breakthroughs in capacity enhancement, as well as improvements in regenerative stability and capacity maintenance. Discussion is made regarding analysis of the affinity of various sorbent formulations for specific components found in typical logistics fuels, and also provided regarding the formulation of layered sorbent beds to optimize performance in high duty cycle applications. Additionally, this paper describes efforts being performed to analyze the effects of fractionated fuels on desulfurization sorbent performance, with the intent to determine means of dealing with inferior grades of fuel, or minimizing system size and complexity, in order to enable further variations of fuel processing equipment, suitable for shipboard and vehicle APU use.

648f Ni/Mh Based Fuel Cell/Battery (FCB) Hybrid System

Bokkyu Choi, Sunmook Lee, Chihiro Fushimi, and Atsushi Tsutsumi. Institute of Industrial Science, The University of Tokyo, 4-6-1 Komaba, Muguro-ku, Tokyo, 153-8505, Japan

Nickel hydroxide/Metal hydride (Ni/MH) is one of the most widely used secondary batteries, e.g. for small portable devices such as cell-phones, MP3 players, and laptop computers, and for large equipments such as robots, electric vehicles (EV) and hybrid electric vehicles (HEV). Recently, there have been active researches on the improvement of conventional Ni/MH secondary batteries to obtain sufficient energy density per weight and power density.

In particular, hybridization of Ni/MH secondary battery with a fuel cell, called Fuel Cell/Battery (FCB), is considered to be a promising method. This can be accomplished by fabricating the positive electrode with the mixture of nickel hydroxide and a small amount of manganese dioxide, a well-known oxygen reduction catalyst. Thus, this hybridized system can function both as a secondary battery and as a fuel cell. In this system, electric power is produced as a battery when it is charged. Furthermore, it can also produce electric power by consuming hydrogen and oxygen gases in a similar way to a fuel cell system when it is discharged. In addition, this system can also function in the following way: Water is electrolyzed when the Ni/MH secondary battery is overcharged, producing hydrogen in the negative electrode and oxygen in the positive electrode, respectively. Thus, the Ni/MH secondary battery can function as a fuel cell by using the stored gases when the battery is overcharged without the supplies of fuels from the outside of the system. Therefore, this leads to improved energy density per weight compared with conventional batteries.

The performances of the Ni/MH based FCB were investigated with cyclic voltammetry (CV) and X-ray Diffraction (XRD). The characteristics of full cell performances were measured with cell-voltage (V) - current density (I) curves. The reaction mechanisms of the positive electrode were also investigated. It was observed that the discharge curves had two plateaus; one at 1.1 V from the reactions of nickel oxyhydroxide, the other at 0.7 V from the fuel cell reactions of manganese dioxide. It was also found that manganese dioxide only functioned as oxygen reduction catalyst, which was confirmed from XRD and CV curves. These experimental results showed the improvement of the overall energy density of the hybrid system compared with conventional batteries.

Key words

Fuel Cell/Battery (FCB) systems, Ni/MH secondary battery, Water electrolysis

690a Microct X-Ray Imaging of Water Movement In a PEM Fuel Cell

Wei Zhu, Zach Dunbar, and Rich Masel. University of Illinois, Urbana-Champaign, 600 South Mathews, Room 204, Urbana, IL 61801

Proper water management has long been identified in the literature as a key component for optimal fuel cell performance. Current literature suggests that the lack of in-situ water data is slowing the progress of water management technology. In the previous literature, water motion has been examined with magnetic resonance imaging (MRI) and neutron imaging(NI), but in this report an entirely new technique was used: microCT. Micro CT uses a high resolution X-ray imaging system to look through the graphite flow channels to provide accurate pictures of water inside of an operating PEM fuel cell. Acquisitions times vary from 2 second to 30 seconds.

The results so far, show qualitative agreement with previous measurements done by MRI and NI, but the resolution is much higher with micro-CT, so many more details can be observed. Water was observed emerging as water droplets on the surface of the diffusion layer. The droplet grew in size, and eventually detached into the flow channels. Drops in the flow channels were found to distort from the expected circular shape before they start to move. Some deviation in the shape of water droplets is observed, when compared to predictions from the modeling literature. This is believed to be due to the role of the channel side walls, as well as possible non-uniformities in the flow channel surface.

690b Direct Carbon Conversion in Carbon Dioxide as the Fuel for Solid-Oxide Fuel Cells

Zongpin Shao, Ran Ran, Rui Cai, and **Xiaohua Lu**. State Key Laboratory of Materials-Oriented Chemical Engineering, Nanjing University of Technology, Xinmofan Road 5#, Nanjing, China

Fuel cells efficiently convert chemical energy to electricity in a silent and environmentally friendly way. They are believed to be a promising alternative power source to traditional mobile and stationary sources, such as the internal combustion engine and coal burning power plants. Among the various kinds of fuel cells, solid-oxide fuel cell (SOFC) has the benefits of the high energy efficiency and fuel flexibility because of their high operating temperature. Therefore, they have received considerable attention in recent years. On the other hand, the remaining huge storage of available coal energy promotes the need for clean utilization of coal. Combining carbon gasification technology and SOFC is a logical step as excess heat from the SOFC process can be used for the coal gasifcation.

In this work, the carbon was directly gasified in CO_2 to CO as the fuel for SOFC. There are several benefits in combining SOFC and gasifier. The coal gasifier supply the SOFC with CO as the fuel, while the SOFC provides heat for the gasifier from the exothermal reactions and CO_2 for the gasification from the anode exhaust. Fe_mO_y oxides were used as the catalysts in the solid carbon conversion reactions. The maximum conversion efficiency of carbon dioxide could be as high as 91 % at 800 ^{o}C compared to only ~60 % without catalyst. An anode-supported fuel cell with the configration of $Ba_{0.5}Sr_{0.5}Co_{0.8}Fe_{0.2}O_{3-¦Ä}$/ $Sm_{0.2}Ce_{0.8}O_{1.9}$/ $Sm_{0.2}Ce_{0.8}O_{1.9}$-Ni was used to test the feasibility of the carbon gasification in CO_2. The power densities of 502, 368 and 216 mW/cm^2 were obtained at 850, 800 and 750 ^{o}C, respectively. The preliminary study indicates that the direct carbon conversion in CO_2 is a promising way to utilization of the solid coal as the fuel for SOFC.

Keywords: Coal, Gasification, Solid-oxide fuel cells.

Acknowledgements

This work was supported by the National Natural Science Foundation of China under contract Nos. 20646002 and 20676061, by National 863 program under contract No. 2007AA05Z133, and by National Basic Research Program of China under contract No. 2007CB209704. Dr Zongping Shao also would like to acknowledge the financial support from Chinese Ministry of Education via the Program for Changjiang Scholars and Innovative Research Team in University (No.IRT0732)

690c Empirical Two-Phase Pressure Drop Correlations and PEM Fuel Cell Cfd Modeling

Sanjiv Kumar, Unit Cell Engineering, Ballard Power Systems, 9000 Glenlyon Parkway, Burnaby, BC V5J 5J8, Canada

Modeling of single-phase flow through pipes and channels of arbitrary cross sections from laminar to turbulent regimes is fairly well understood and Computational Fluid Dynamics (CFD) is commonly employed to predict pressure drops and other flow behaviors. Several attempts have been made by many researchers to develop CFD based predictive tools to describe multi-phase flows with varying degree of simplification and a companion paper on the subject presents a two fluid model approach in the context of Fuel Cells.

Still a commonly employed engineering approach is to use Lockhart & Martinelli type of empirical correlations in two-phase fluid flow. Straightforward applicability of such empirical correlations for fuel cell application is much more doubtful as liquid water is added along the length of the channel due to electrochemical reaction and typically some estimate of liquid water present in the flow channels is required to correctly describe the pressure drop. In as much as this is a good engineering simplification the flow behavior in fuel cells is different from the classical two-phase flow where an incoming quality defines the second phase entering the domain, which is the starting point of such empirical correlations.

In this paper empirical correlations based on the experimental data for different mini-channels used in PEM fuel cell cathode flow fields are described. These correlations are used for describing the two-phase flow pressure drops in cathode flow fields of PEM fuel cells. These empirical correlations are also compared with Lockhart & Martinelli type models based on assumed liquid volume in channels. The developed correlations serve as a good water management design tool for the fuel cell designers.

Furthermore, in absence of such empirical correlations, it is not possible to account for significant pressure drops in the cathode channel where liquid water, that is the product of the electro-chemical reaction as well as due to other water transport mechanisms, is present in the flow channel and significantly influences the pressure drop. Typically in some applications, because of the liquid water, the pressure drops can be three times higher than that based on laminar flow consideration alone and this cannot be predicted by a single-phase CFD model in absence of a "laminar like" friction factor in the model. In addition to posing the water management problem for PEM fuel cell flow fields, this significantly impacts the concentration profile of reactants along the channel and the CFD based fuel cell model must be able to account for such pressure variations. In this paper the developed correlation is used in a CFD based fuel cell model and predicted pressure drops and performance of fuel cell are compared with experimental data.

Lattice-Boltzmann Simulations of Multiphase Flows in Gas-Diffusion-Layer (GDL) of a PEM Fuel Cell

Shiladitya Mukherjee[a], J. Vernon Cole[a], Kunal Jain[b], and Ashok Gidwani[a]

[a] CFD Research Corporation, Huntsville, Alabama 35805, USA
[b] ESI US R&D Inc, Huntsville, Alabama 35806, USA

Introduction

Improved power density and freeze-thaw durability in automotive applications of Proton Exchange Membrane Fuel Cells (PEMFCs) requires effective water management at the membrane. This is controlled by a porous hydrophobic gas-diffusion-layer (GDL) inserted between the membrane catalyst layer and the gas reactant channels. The GDL distributes the incoming gaseous reactants on the catalyst surface and removes excess water by capillary action. There is, however, limited understanding of the multiphase, multi-component transport of liquid water, vapor and gaseous reactants within these porous materials. This is due primarily to the challenges of in-situ diagnostics for such thin (200 – 300 μ), optically opaque (graphite) materials. Transport is typically analyzed by fitting Darcy's Law type expressions for permeability, in conjunction with capillary pressure relations based on formulations derived for media such as soils. Therefore, there is significant interest in developing predictive models for transport in GDLs and related porous media. Such models could be applied to analyze and optimize systems based on the interactions between cell design, materials, and operating conditions, and could also be applied to evaluating material design concepts.

Recently, the Lattice Boltzmann Method (LBM) has emerged as an effective tool in modeling multiphase flows in general [1-3], and flows through porous media in particular [4-6]. This method is based on the solution of a discrete form of the well-known Boltzmann Transport Equation (BTE) for molecular distribution, tailored to recover the continuum Navier-Stokes flow [7,8]. The kinetic theory basis of the method allows simple implementation of molecular forces responsible for liquid-gas phase separation and capillary effects. The solution advances by a streaming and collision type algorithm that makes it suitable to implement for domains with complex boundaries. We have developed both single and multiphase LB models and applied them to simulate flow through porous GDL materials. We will present an overview of the methods as implemented, verification studies for both microstructure reconstruction and transport simulations, and application to single- and two-phase transport in GDL structures. The applications studies are designed to both improve understanding of transport within a given structure, and to investigate possible routes for improving material properties through microstructure design.

Numerical Model

In this section, we describe implementation of a multi-dimensional, multi-phase Lattice-Boltzmann Model (LBM) with a model for surface wettability, and the reconstruction scheme for the porous gas-diffusion-layer (GDL).

The Lattice-Boltzmann Model (LBM)

The LBM is based on well-known Boltzmann Transport Equation (BTE) given by,

$$\frac{Df(\mathbf{x},\xi,t)}{Dt} = \frac{\partial f}{\partial t} + \xi.\nabla f + \mathbf{F}.\nabla_{\xi} f = \left(\frac{\partial f}{\partial t}\right)_{coll} \qquad [1]$$

Here, f is the probability distribution function describing distribution of particle population over velocities ξ, at location $\mathbf{x}$ at time t and $\mathbf{F}$ is the mean-molecular interaction force. The right side of the equation is the rate of change in f due to intermolecular collisions. f is related to the local density ρ and momentum $\rho\mathbf{u}$ as,

$$\rho(\mathbf{x},t) = \int f(\mathbf{x},\xi,t)d\xi \qquad [2]$$

$$\rho(\mathbf{x},t)\mathbf{u}(\mathbf{x},t) = \int f(\mathbf{x},\xi,t)\xi d\xi \qquad [3]$$

In LBM, the collision operator is modeled by Bhatanagar-Gross-Krook (BGK) model [9],

$$\left(\frac{\partial f}{\partial t}\right)_{coll} = -\frac{f - f^{eq}}{\lambda} \qquad [4]$$

The Boltzmann model assumes that particle distributions approach local equilibrium linearly over a characteristic time λ. This characteristic time is related to fluid viscosity ν. While simple, this model has well-known limitations in applications such as multiphase and porous media flow. In particular, for porous media flow, employing the BGK model results in viscosity dependence of the computed permeability. We have also reconfirmed this observation through our own set of simulations. Hence, in this work, a multiple-relaxation-time (MRT) collision model is implemented [8,10]. In the MRT model, the single relaxation time λ is replaced by a collision-matrix given by,

$$\left(\frac{\partial f_{\alpha}}{\partial t}\right)_{coll} = -\Lambda_{\alpha\beta}(f_{\beta} - f_{\beta}^{\ eq}) \qquad [5]$$

Here, α, β corresponds to particle velocities ξ_{α} and ξ_{β}, respectively. Employing the collision-matrix allows separation of relaxation time-scales between hydrodynamic modes such as velocity, pressure and stress-tensors, thus improving numerical accuracy and stability.

Multiphase LBM

In the multiphase model of He et al. [1], the mean-molecular interaction force $\mathbf{F}$ is formulated as sum of a phase segregation and surface tension force. The phase segregation force is expressed as the gradient of the non-ideal part of the equation-of-states (EOS), which separates the fluid into respective liquid and gas phase densities by entropy minimization. The surface tension force depends on the interfacial curvature, and is scaled by the strength of the molecular interaction.

In numerical implementation, two separate distribution functions are used. One distribution function f is for a phase-tracking variable, known as the index-function φ. The transport equation of f

recovers an advection-diffusion equation known as the Cahn-Hilliard equation. Transition of φ from the limiting values at the liquid to gas phase occurs over finite number of lattice nodes. The interface is implicitly defined as the contour of φ having the average of these limiting values. Fluid properties, such as, density and viscosity are interpolated from φ. A second distribution function g is used to compute pressure and momentum. The transport equation of g recovers the Navier-stokes equation with a surface tension term. The surface tension is computed from φ as proportional to $\varphi\nabla\nabla^2\varphi$. In this work, the first and second-order derivatives are computed as weighted sum of the second-order central difference along the lattice-velocity directions. The limiting values of φ are obtained analytically from the EOS and can be further refined from the simulation of liquid film or drop equilibration.

Surface Wettability

Mukherjee et al. [2] modeled wettability by adding an external force near the surface, mimicking molecular attraction/repulsion between liquid drop molecules and solid surface molecules as,

$$\mathbf{F}_w = -K_w \varphi_l \mathbf{n}_s \qquad [6]$$

Here, K_w is the strength of interaction parameter and $\mathbf{n}_s$ is the surface normal. In this work, we employ a similar model by Yiotis et al. [11] for its simplicity in implementing on complex boundaries, such as, randomly oriented fibers. The model assigns a φ value to the solid, between the liquid and gas limits to compute surface tension forces near the solid boundary.

Microstructure Reconstruction

The fiber morphology of Toray and SGL non-woven carbon papers are generated as continuous cylinders of fixed diameter f_d, with their axis along parallel planes separated by a spacing of f_d. Within a plane the fibers are randomly orientated and may intersect each other. This approach is similar to that reported by Schulz et al. [4].

Results and Discussions

Single-phase Gas Permeability

The single phase LB model is benchmarked against the test cases of decaying Taylor Vortices and pressure-driven channel flow. The model is then used to simulate gas flow through numerically reconstructed porous material and compute absolute permeability. Figure 1 shows an idealized porous structure of simple-cubic (SC) arrangement of solid spheres. The permeability is computed from the average flow velocity u by applying Darcy's law,

$$u = \frac{\kappa}{\mu\varepsilon}\left(\frac{P_{in} - P_{out}}{L}\right) \qquad [7]$$

Here, κ is the permeability, ε the porosity, μ the viscosity, P the pressure, and L the channel length. In Table 1, computed permeability is compared with the analytical solution of Chapman and Higdon [12] at different porosity by varying the sphere size. The values are in agreement within 5% of the analytical solution.

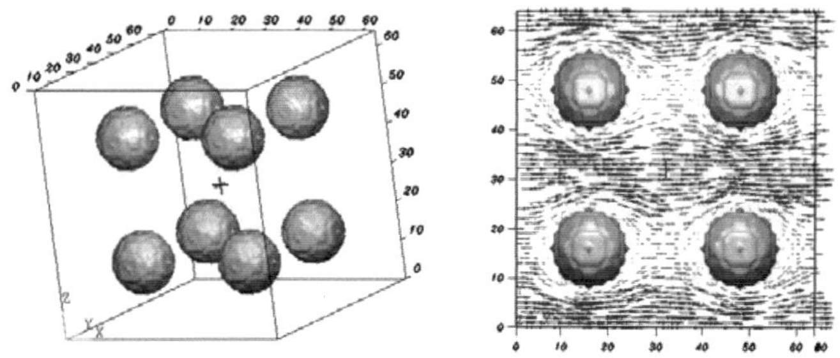

(a) Solid spheres

(b) Flow around spheres

TABLE 1. Non-Dimensional Permeability of Simple-Cubic (SC) Arrangement of Spheres.

Porosity, ε	**Analytical, κ/d²**	**Computed, κ/d²**	**% Error**
0.992	0.2805	0.2843	1.3
0.935	0.0761	0.0749	1.6
0.779	0.0192	0.0189	1.6
0.477	0.0025	0.0026	4

Figure 1. Flow through simple-cubic arrangement of solid spheres.

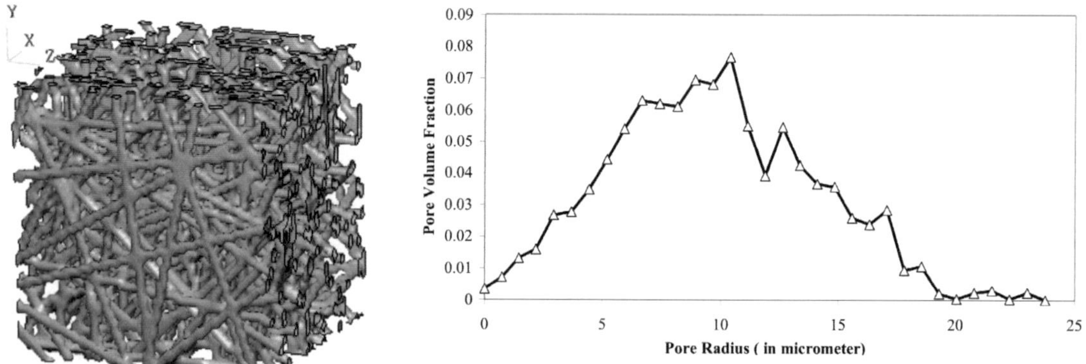

Carbon paper

Pore size distribution

Figure 2. Numerically reconstructed carbon paper and its pore size distribution. The porosity is 0.78, comparable to commercially available Toray090.

TABLE 2. Carbon-Paper Microstructure Permeability.

Material	**Measured**	**Simulation**	**% Error**
Toray090 (Through-plane)	8.3	6.241	-24.8
Toray090 (In-plane)	-	8.647	-
SGL10BA (Through-plane)	18	21.71	20.6
SGL10BA (In-plane)	33	30	9.1

Next, in-plane and through-plane permeability are computed for Toray090 and SGL10BA with porosity 0.78 and 0.88, respectively. In Figure 2 numerically generated microstructure for Toray090 is shown. The reconstructed GDL is 200μm thick with spatial resolution of 3.4μm. In Table 2, the computed and measured values of κ are compared. The values are in the unit of *darcy* (10^{-12} m^2). SGL10BA has higher permeability than Toray090 due to higher porosity. The in-plane permeability is greater than through-plane value. The difference is due to the preferred fiber orientation in the material plane and suggests that alignment to the mean-flow direction have less resistance to the flow. The computed trends are consistent with these observations and are similar to those reported in the numerical work of Schulz et al. [4].

Influence of fiber orientation on permeability is investigated next. Figure 3 shows microstructures when the fibers are constrained to orient within an angle of φ with the y-axes in the in-plane. Permeability is plotted in Figure 6. The κ_{th} decreases with decreasing φ, however, the change is rather small. As example, κ_{th} at $\varphi = 5°$ and 90° are within 15%. The effect is more pronounced in the in-plane direction. Here, $\kappa_{in,z}$ increases by about 75% and $\kappa_{in,y}$ decreases by 47% as φ is lowered from 90° to 5°. The results suggest that the in-plane permeability can be increased significantly along a preferred direction by orienting the fibers along it, without resulting in significant drop of through-plane permeability. This information can be used to enhance gas-diffusion in GDL. The GDL-material can be constructed in two layers. The gas-channel side consisting of fibers with preferred orientations, assisting in-plane gas flow between neighboring gas-channels, and the catalyst side having fibers randomly oriented thereby allowing uniform distribution of incoming gas on the reaction sites.

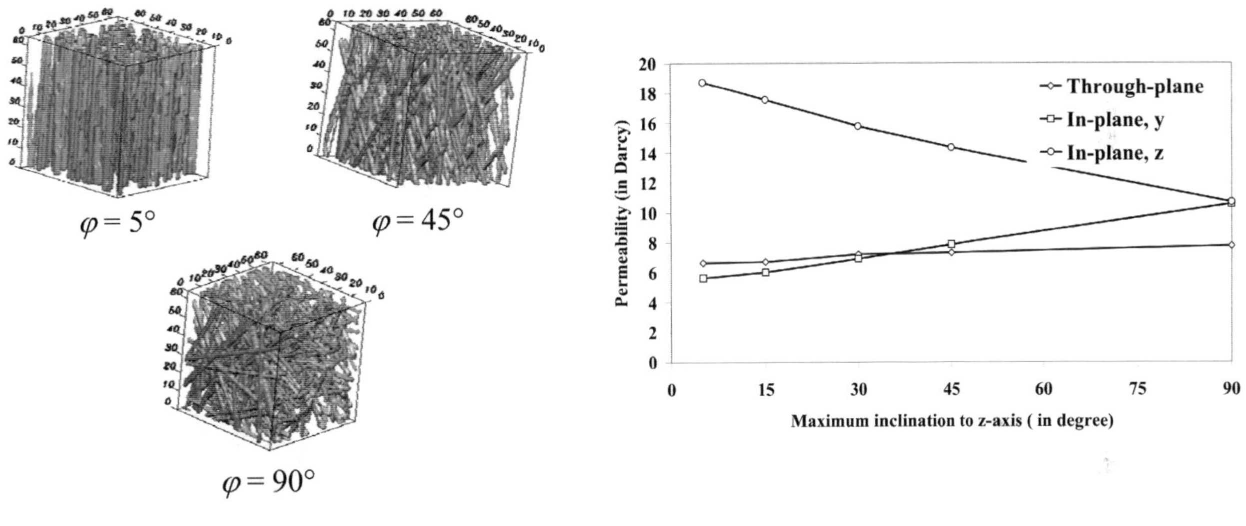

Figure 3. Through and In – plane permeability at various fiber-orientations.

Porous microstructure under external compression is modeled following the same approach as by Schulz et al. [4] i.e. by linearly scaling the height of the solid voxels from the base of the material. Figure 4 shows Toray090 microstructure at $c = 0.6$ and 0.8 and through-plane gas-permeability at different compression ratios for Toray090 and SGL10BA. The measured data is by Dohle et al. [13] on SGL10BA. At higher compression, the GDL thickness decreases but there is no change in the fiber volume. Therefore, porosity and permeability decreases. Computed and measured data are consistent with this expectation.

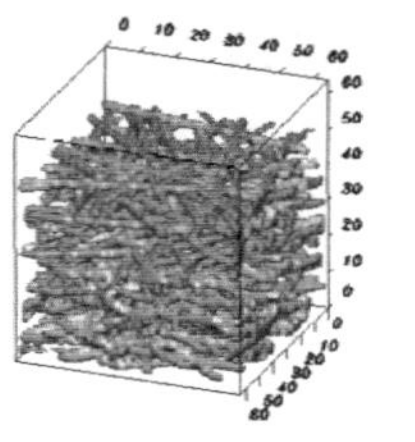

Toray090, c = 0.8

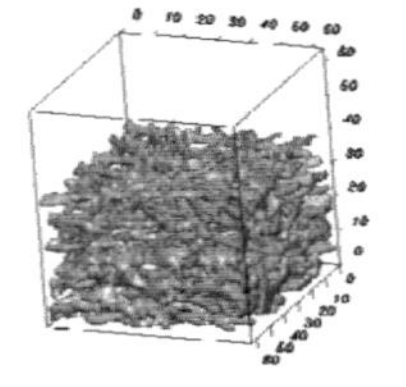

Toray090, c = 0.6

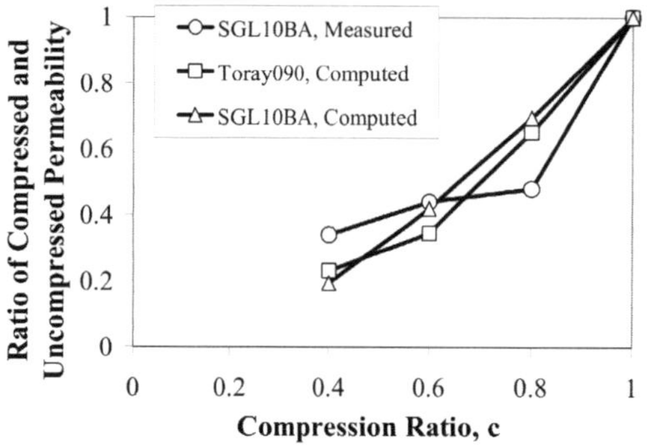

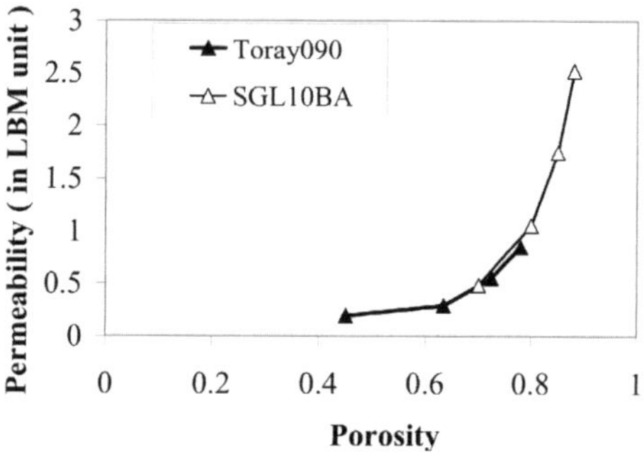

Figure 4. Permeability of Toray090 and SGL10BA under compression.

Multiphase Transport - Capillary Effect

The surface wettability model is validated by simulating drop equilibration on a partially wetted surface. Initially the drop is set to a hemispherical shape corresponding to the contact angle of 90°. As it equilibrates the contact angle may differ depending on the wetting/non-wetting property of the surface. In Figure 5 shape of the equilibrium drop is shown at various wettability including hydrophilic, hydrophobic and neutral.

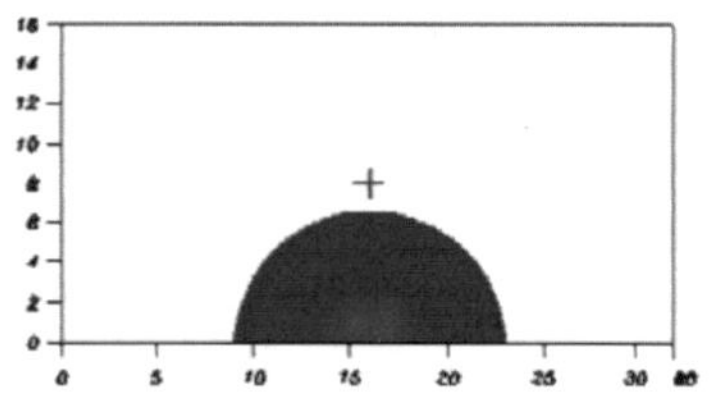

(a) Neutral

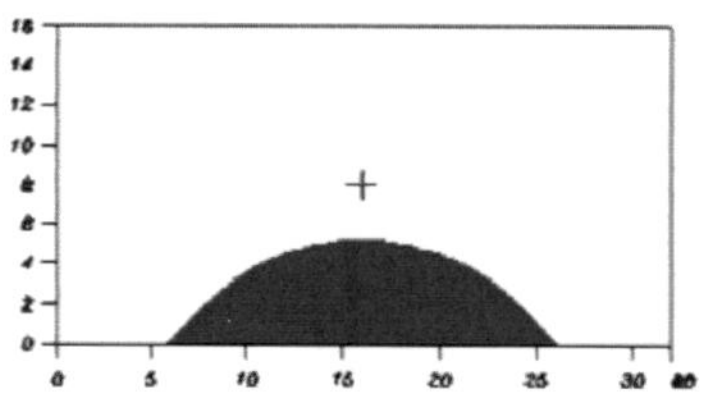

(b) Hydrophilic

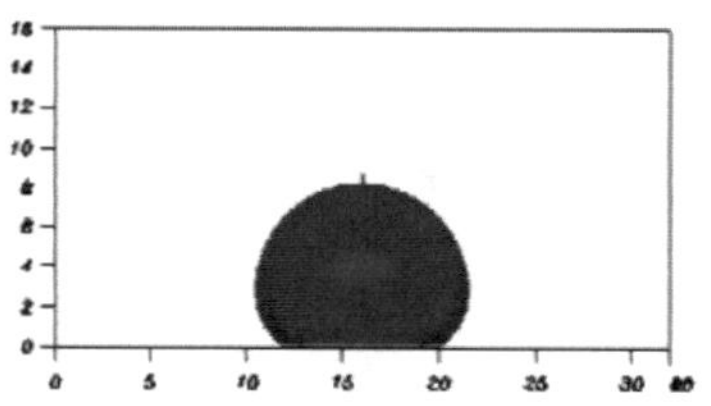

(c) Hydrophobic

Figure 5. Drop in equilibrium on partially wetting surfaces.

Figure 6 shows liquid drainage and imbibitions due to capillary action. Initially a tube with circular cross-section is half filled with liquid. The inlet and outlet pressures are set as equal. As the wetting property of the tube is set to hydrophobic, neutral or hydrophilic the liquid-gas interface moves due to capillary pressure. The liquid drains out of the tube when θ_{eq}>90°, fills in the tube when θ_{eq}<90°, and remains stationary when θ_{eq}=90°.

Parallel flow of liquid and gas in a channel is shown in Figure 7. The flow is driven by an imposed pressure gradient between inlet and outlet. The liquid to gas density and viscosity ratios are set to 10. As the flow evolves the liquid-gas interface stays parallel to the flow direction, however, velocity in the gas is greater than the liquid velocity due to lower density and viscosity.

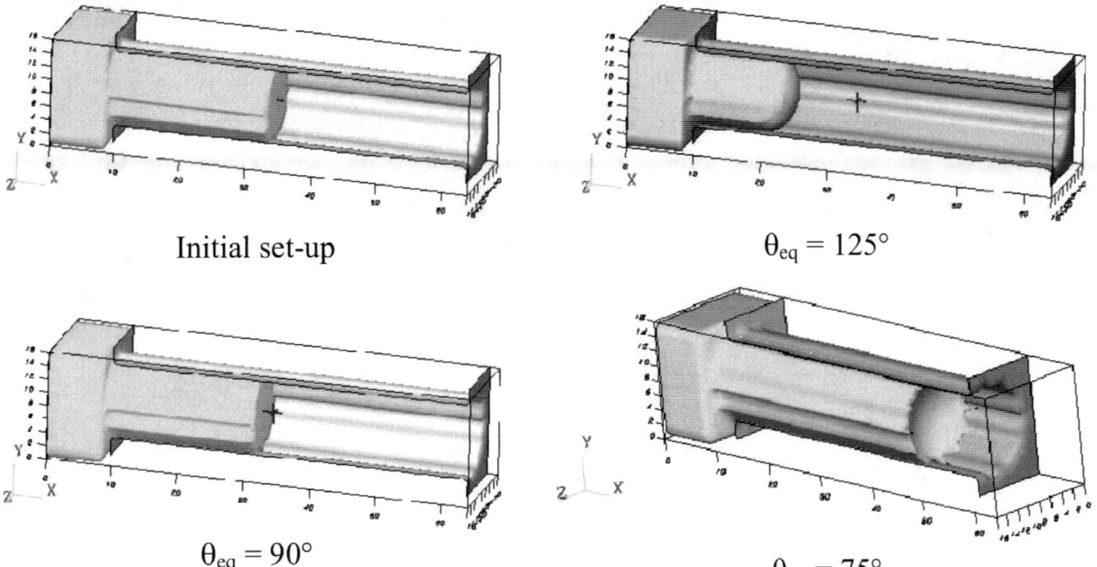

Initial set-up $\theta_{eq} = 125°$

$\theta_{eq} = 90°$ $\theta_{eq} = 75°$

Figure 6. Liquid drainage and imbibition due to capillary action in a tube.

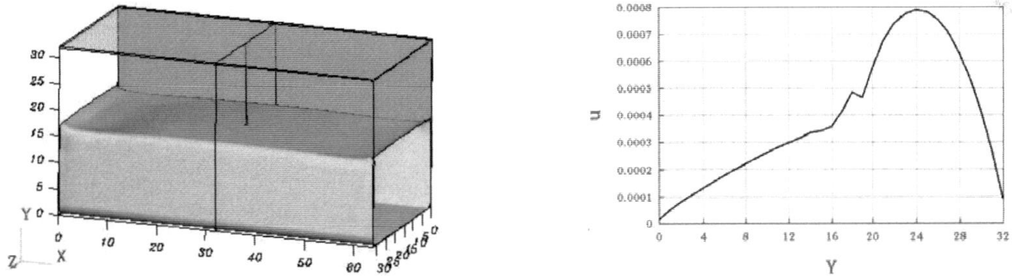

Parallel flow of liquid and gas   Velocity profile in liquid and gas

Figure 7. Separated flow of liquid and gas in a channel driven by pressure difference between inlet and outlet. The liquid to gas viscosity and density ratios are 10.

In Figure 8, we consider flow through a GDL. The computational domain is resolved with 93 nodes in the flow direction and 64×64 nodes at the plane normal to it. Unit node spacing corresponds to 3.4μm. The GDL is 200μm thick comprising of 25 layers of carbon fibers of diameter 8μm, the in-plane area is 218μm×218μm and porosity 0.78. An open space of thickness 109μm is provided at the gas-channel side to allow liquid-breakthrough and formation of multiple droplets. Liquid is introduced at the boundary on the catalyst side. The side-boundaries are no-slip walls of neutral wettability i.e. $\theta_{eq} = 90^\circ$. In Figure 8, snap-shots of liquid-breakthrough are shown at different fiber hydrophobicity. Liquid breaks through the channel side of the GDL in the form of droplets. These droplets may coalesce forming larger drops. As the hydrophobicity is increased the saturation at the break-through decreases. Liquid flows as small streams leading to surface pores. These observations are consistent with other published data [14-16].

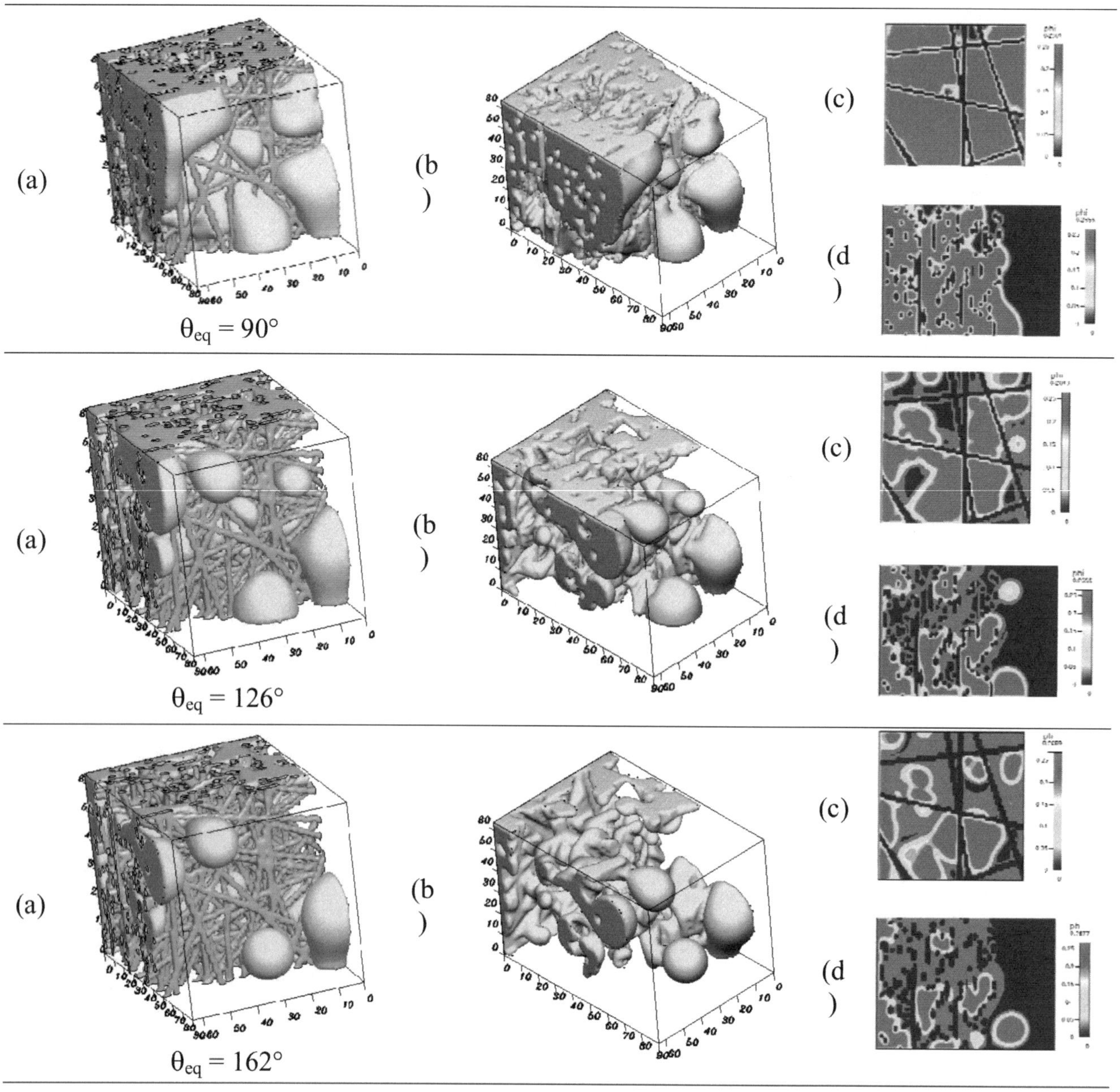

Figure 8. Influence of fiber wettability on liquid saturation as it breaks through the porous gas-diffusion layer (GDL). (a) Emergence of liquid drops at the GDL and gas-channel interface. (b) Liquid distribution inside the porous medium. (c) Liquid-gas distribution on a plane normal to the flow. (d) Liquid-gas distribution on a plane parallel to the flow. Liquid saturation decreases as fiber hydrophobicity increases.

Conclusions

In this work, the lattice-Boltzmann method (LBM) is demonstrated as an effective numerical tool in modeling single and multiphase transport through porous microstructures representing the gas-diffusion-layer of a PEM Fuel Cell. The model is suitable to capture effects of fiber orientation, external compression, and capillary pressure for multiphase transport. Flows through carbon paper

GDL show emergence of liquid droplets from the gas-channel side. The fiber hydrophobicity has strong influence on the liquid flow paths inside the GDL. As hydrophobicity increases liquid flows as separated streams, resulting in a low liquid saturation at break through.

Acknowledgments

This material is based upon work supported by the Department of Energy under award number DE-FG36-07G017010.

Disclaimer

This report was prepared as and account of work sponsored by an agency of the United States Government. Neither the United States Government nor any agency thereof, nor any of their employees, makes any warranty, express or implied, or assumes any legal liability or responsibility for the accuracy, completeness, or usefulness of any information, apparatus, product, or process disclosed, or represents that its use would not infringe privately owned rights. Reference herein to any specific commercial product, process, or service by trade name, trademark, manufacturer, or otherwise does not necessarily constitute or imply its endorsement, recommendation, or favoring by the United States Government or any agency thereof. The views and opinions of authors expressed herein do not necessarily state or reflect those of the United States Government or any agency thereof.

References

1. He X., Chen, S. and Zhang, R. (1999), *J. Comp. Phys.*, 152, 642.
2. Mukherjee S. and Abraham, J. (2007), *J. Coll. Inter. Sci.*, 312, 341.
3. Mukherjee S. and Abraham, J. (2007), *Phys. Fluids*, 19, 052103.
4. Mukherjee,S., Cole, J.V., Jain, K., and Gidwani, A. (2008), 214th Meeting of Electrochemical Society, Honolulu, HI, Oct. 12-17.
5. Schulz, V. P., Becker, J., Wiegmann, A., Mukherjee, P.P., and Wang, C.Y. (2007), *J. Electrochem. Soc.,* 154, B419.
6. Niu, X. D., Munekata, T., Hyodo, S.A. and Suga, K. (2007), *J. Power Sources*, 172, 542.
7. Mukherjee S. and Abraham, J. (2007), *Phys. Rev. E*, 75, 026701.
8. Mukherjee S. and Abraham, J. (2007), *Computers Fluids*, 36, 1149.
9. Bhatnagar, P., Gross, E. and Krook, M. (1954), *Phys. Rev.*, 94, 511.
10. D'Humie'res, D., Ginzburg, I., Krafczyk, M., Lallemand, P., and Luo, L.S. (2002) *Phil. Trans. Roy. Soc. Lon. A*, 360, 437.
11. Yiotis, A.G., Psihogios, J., Kainourgiakis, M. E., Papaioannou, A., and Stubos, A.K. (2007), *Coll. Surf. A: Physiochem. Eng. Aspects*, 300, 35.
12. Martys, N.S. and Hagedorn, J.G. (2002), *Materials and Structures*, 35, 650.
13. Dohle, H., Jung, R., Kimiaie, N., Mergel, J., and Müller, M. (2003), *J. Power Sources*, 124, 371.
14. Lister, S., Sinton, D. and Djilali, N. (2006), *Journal of Power Sources*, 154, 95-105.
15. Bazylak, A., Sinton, D., Liu, Z.-S., and Djilali, N. (2007), *Journal of Power Sources*, 163, 784-792.
16. Bazylak, A., Sinton, D., and Djilali, N. (2008), *Journal of Power Sources*, 176, 240-246.

690e Development of a Tubular Proton Exchange Membrane Fuel Cell (PEMFC) –Experimental and Modeling Studies

Brian Bullecks[1], Debangsu Bhattacharyya[1], Raghunathan Rengaswamy[2], and Gregory A. Campbell[1]. (1) Chemical Engg. Dept., Clarkson University, 8 Clarkson Avenue, Potsdam, NY 13699, (2) Chemical Engineering, Clarkson University, Box 5705, Clarkson University, Potsdam, NY 13699-5705

PEM fuel cells (PEMFC) are being strongly considered for portable applications because of its low operating temperature, quick start up and high current density. However, significant reduction in cost is required to make this technology commercially viable. A reduction in the cell hardware can help in substantial reduction of the cost of a cell. An increase in the power/weight ratio can reduce the parasitic losses for automotive applications of the cell.

With these objectives in mind, a tubular PEMFC is developed. Typical PEM fuel cells are planar. Although the tubular geometry has been successfully used in other type of fuel cells such as solid oxide fuel cell (SOFC), it is not common in the case of PEMFC. The tubular cell is observed to have a higher current density than a planar cell using the same commercial 5 layer MEA and under the same operating conditions. A significant reduction in the hardware cost is achieved. The power/weight ratio is found to be several times higher than the commercial planar PEMFCs. Possible reasons for such improvements are investigated through EIS studies. Analysis of this study along with a study on the life expectancy of this cell will be presented. The problems, identified in the process of evaluation, and their possible remedial measures will also be discussed.

In order to optimize the cell dimensions, a steady state model of the cell is developed. In this isothermal half cell model, mass and momentum conservation equations are considered inside the cathode gas flow channel. Mass/species conservation equations are written for the gas diffusion layer and the reaction layer. The reaction layer is characterized as both macro homogenous and spherical agglomerate models. Effects of these characterizations on the model predictions will be presented. In order to reduce the time for computations, the method of Thiele Modulus and effectiveness factor are adopted for the spherical agglomerate characterization. In this model, liquid water is considered in all the layers of the cell. The evaporation/condensation process for water is also considered. Interfacial liquid water saturation can be strongly influenced by the surface properties and the operating conditions of the cell. In order to assess the effects of different surfaces, a sensitivity study is done based on the boundary conditions of the liquid water saturation. The model is validated by the experimental data collected over a wide range of air flow rate and operating pressure.

A NOVEL METHOD FOR THE PERFORMANCE ANALYSIS OF A MOLTEN CARBONATE FUEL CELL

Choong-Gon Lee[1], Hee-Chun Lim[2], Do-Hyung Kim[2]*
[1]Department of Chemical Engineering, Hanbat National University
San 16-1, Dukmyung-dong, Yuseong-gu, Daejeon, 305-719 Korea
Email: leecg@hanbat.ac.kr
[2]Korea Electric Power Research Institute (KEPRI),
103-16, Munji-dong, Yusong-ku, Taejon, 305-380 Korea

Abstract. This work has been focused on the behavior of overpotential increase by utilization rise in a molten carbonate fuel cell. The behavior is generally explained by Nernst loss, which is a kind of voltage loss due to the thermodynamic potential gradients at a polarization state by the concentration distribution of reactant species through the gas flow direction. The evaluation of Nernst loss was carried out with traditional experimental way of constant gas utilization (CU). On the other hand, overvoltage due to the gas-phase mass-transport resistance at the anode and cathode showed dependence on the utilization which was measured with the inert gas step addition (ISA) method. Since the Nernst loss was assumed by the thermodynamic reasons, the voltage loss can be calculated by the Nernst equation, calling simple calculation (SC) in this work. The three values of voltage loss by CU, ISA, and SC were compared, showing these values rise with utilization increase within acceptable deviations. When we consider that the anode and cathode reactions are significantly affected by the gas-phase mass transfer, the behavior strongly implies that the voltage loss is attributed not to the thermodynamic reasons, i.e, Nernst loss, but to the kinetic reason of mass-transfer resistance in the gas phase.

Keywords: MCFC; overpotential; mass transfer; Nernst loss

1. Introduction

Understanding of the reaction kinetics in the molten carbonate fuel cell (MCFC) is required for its practical development because the performance is determined by the reaction kinetics at the electrodes. Since the electrode kinetics determines performance via overpotential at the anode and cathode, the correct evaluation of overpotential provides clues for the analysis of reaction characteristics. Voltage loss in the MCFC generally comprises of ohmic loss (η_{IR}) and reaction overpotential at the anode (η_{an}) and cathode (η_{ca}); ohmic resistance is due to the electrical resistance among the cell components and reaction overpotential is ascribed to the resistance in the electrochemical reactions at both electrodes. Thus, output voltage (V) can be expressed as follows;

$$V = E_{OCV} - \eta_{IR} - (\eta_{an} + \eta_{ca}) \tag{1}$$

where, $E_{OCV} = E_0 + \frac{RT}{2F}\ln\left(\frac{[H_2]}{[H_2O][CO_2]_{an}}[O_2]^{0.5}[CO_2]_{ca}\right)$, and E_0 is the standard potential. Subscripts an and ca represent anode and cathode, respectively. Other symbols have their own meanings.

The MCFC, normally running around 650°C with carbonate electrolyte, has relatively severe operating conditions; 1) it uses porous electrode with corrosive carbonate electrolytes melting around 500°C, 2) combination of cell components such as electrodes, electrolyte, and matrices may have deviations by the cells, 3) the cell is assembled with thinner and lighter components, not allowing stable installation of reference electrode to analyze overpotential at the electrodes. The complete analysis of the overpotential and reaction kinetics has been obstructed by the reasons.

In the previous papers, the reaction characteristics have been illuminated in the MCFC with several experimental ways; inert gas step addition (ISA) method [Lee et al. 2003], reactant gas addition (RA) method [Lee et al. 2003] and water head (WH) measurements [Lee et al. 2004]. They mostly treated gas flow effects on the overpotential, indicating consistently that the anode is principally mass-transfer controlling process in the gas phase, whereas the cathode is mixed mass-transfer controlling processes in the gas and liquid phases.

As a voltage loss in the MCFC, Nernst loss has been suggested [Sampath et al. 1980]. It is based on the assumption that concentration distribution occurs along the gas flow way at a polarization state, and it induces potential gradient in the cell according to the Nernst equation. Then, the potential difference between gas inlet and outlet behaves as overvoltage, called Nernst loss. However, the thermodynamic relation of Nernst equation is based on the equilibrium state which the concentrations of reactants are the same everywhere in the cell. It has been reported that MCFC in a polarization state has concentration distribution vertically from the gas phase to the electrode surface [Lee et al. 2003]. Then it is questionable that the potential calculated from the bulk concentration can represent electrode potential while the concentration on the electrode surface may have different value from the bulk due to the concentration gradient at polarization state.

The dependence of overpotential on the utilization at both electrodes was illuminated by comparing the Nernst loss from the traditional ways with overpotential from a novel experimental way. As the conventional ways, the constant gas utilization (CU) method and simple calculation (SC) were employed. Those are considered as experimental and theoretical methods to obtain the Nernst loss. As a novel way, inert gas step addition (ISA) method was used. Voltage losses were obtained from CU, SC, and ISA and compared among them at various anode and cathode utilizations.

2. Experimental

The MCFC used in this work had about 100 cm^2 of geometrical surface area of electrode. Anode was a porous slab of Ni-Al alloy while the cathode was a porous Ni slab. Matrix was fiber reinforced $LiAlO_2$, and the electrolyte was (62+38)mol% $(Li+K)_2CO_3$ eutectics. The single cell was operated at an atmospheric condition, and temperature was kept at 923K.

More details of the cell components and cell operation were described in a previous work [Lee et al. 2003].

Mixtures of 80 mol% H_2 and 20 mol% CO_2 through the 53°C water bubbler was supplied to the anode, whereas mixtures of 70 mol% air and 30 mol% CO_2 was fed to the cathode. Utilization, a ratio of consumed gas amount to feed amount, of anode (u_f) and cathode (u_{ox}) ranged from 0.2 to 0.6 at 15 A. The gas flows were controlled by several MFCs (Mass Flow Controllers).

Steady state polarization has been carried out to obtain the effect of utilization change on the overvoltage. The current was raised from 0 to 15 A with a galvanostat at a fixed reactant flow rate. Then, we could obtain total overvoltage with respect to the utilization.

As a conventional way to measure Nernst loss, constant gas utilization (CU) has been employed. Figure 1 shows how to measure the Nernst loss at the anode ($\Delta V_{u,an}$) and cathode ($\Delta V_{u,ca}$) with CU method. Since the Nernst loss is a function of utilization, a difference between OCV(E_{OCV}) and an extrapolated voltage to 0 current from the current-voltage line of constant utilization is regarded as Nernst loss at the utilization.

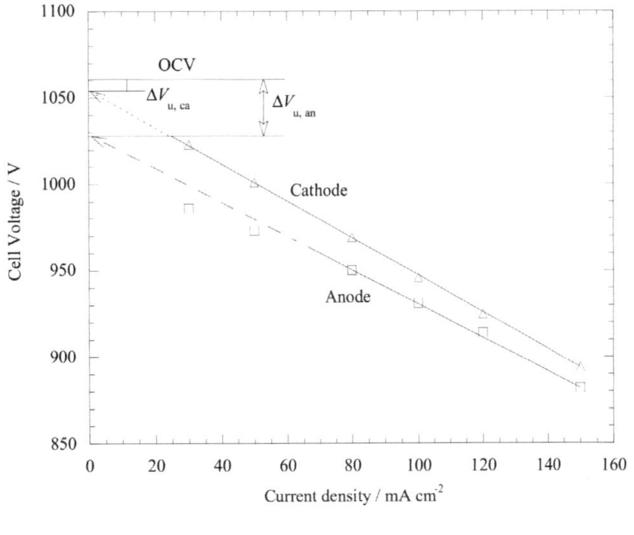

Fig. 1 Estimation of voltage loss with constant gas utilization (CU) method at the anode (u_f=0.2) and cathode (u_{ox}=0.4), 923K, 1 atm.

$$\Delta V_u = E_{OCV} - V(u, i \to 0) \qquad (2)$$

The constant utilization was kept by controlling the gas flow rate according to the applied currents; thus the gas flow rates were proportionally reduced to the applied currents. The value of $\Delta V_{u,an}$ is much larger than $\Delta V_{u,ca}$ as shown in Fig.1. Thus it is known that anode has much larger Nernst loss than cathode.

As a novel method, inert gas step addition (ISA) method was employed in this work. ISA adds certain flow rate of inert gas to the cell. The added inert gas raises reactant gas flow rate without changing gas composition while the inert gas arrives at the cell. Then, overvoltage change by the flow rate shift can be obtained by measuring voltage shift. Thus, ISA can provide overvoltage with respect to the reactant flow rate and utilization. Nitrogen gas was used as an inert gas to the anode and cathode, and its flow rate ranged from 0.1 to 1.0 L min^{-1}. Addition port for N_2 was installed about 4 m before the anode and about 1.5 m ahead of the cathode. All the gas tubes are 1/4 inches of outer diameter with stainless steel. More detailed characteristics of ISA method were described in a previous paper [Lee et al. 2003].

3. Results and discussions

Figure 2 shows results of steady state polarization (SSP) with respect to the anode (a) and cathode utilization (b). Since the utilization is a ratio of consumed reactant amounts to fed amount at a fixed current load, higher utilization means lower flow rate of reactant gases. The clear dependence of voltage on the utilization and flow rate is observed. The SSP represents total overvoltage by the difference between output voltage (V) and open circuit voltage (E_{OCV}), then it is obvious that higher utilization has larger overvoltage. In addition, the distinguishable change of the slope at the anode indicates that the anodic overpotential more strongly depends on the flow rate than the cathodic one. The electric conductivity of carbonate electrolyte is generally seldom dependent on the flow rates, thus the flow rate change is wholly responsible for the overpotential variation.

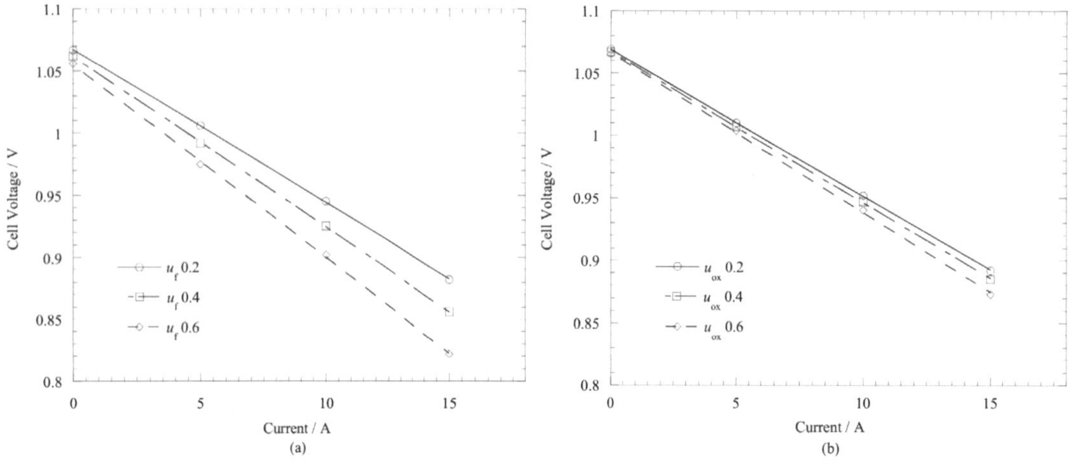

Fig. 2. Results of steady state polarization (a) at anode of u_{ox}=0.4 (15A), (b) at cathode of u_f=0.2 (15A), 923K, 1atm.

In a previous work of Sampath et al. [Sampath et al. 1980], the behavior has been interpreted thermodynamically. Since the cell consumes a certain amount of gas at a fixed current load, flow rate change leads to various concentration distributions along the gas flow way. More severe concentration distribution is expected at lower flow rate. From a theoretical view point [Bard et al., 2001], potential of a reversible system is determined by the concentration of reactant and product species. Thus the concentration distribution gives rise to potential gradient along the gas flow way. Steeper potential gradient occurs at lower gas flow rate and higher utilization at a fixed current. They suggested that the potential gradient behaves as a voltage loss, called Nernst loss. Thus larger Nernst loss is suggested at higher utilization and lower flow rate. According to their suggestion, the large overvoltage at a high utilization is due to the significant amount of Nernst loss.

However, Lee et al. [Lee et al. 2003] has suggested that MCFC has significantly high reaction rate at the anode and cathode, thus the reactions at the anode are mass-transfer controlling processes in the gas phase while the cathode reaction is mixed mass-transfer limiting processes in the gas and liquid phases from the ISA method. They also reported that the anode overpotential is mostly attributed to the mass-transfer resistance of CO_2 and H_2O at a normal operating condition (H_2:CO_2:H_2O=0.69:0.17:0.14 atm) [Lee et al. 2005].

Figure 3 shows results of ISA measurement with addition of 0.3 L min^{-1} N_2 to the anode (a) and cathode (b). As reported in a previous work [Lee et al. 2003], N_2 addition to the anode immediately raises voltage because the addition enhances reactant flow rate without changing composition. Thus the voltage gain ($\Delta V_{I,an}$) is obtained by the overvoltage shift due to the flow rate variation. Since the overpotential induced by the gas-phase transport is a function of flow velocity and the flow rate at a fixed current is a function of utilization, the voltage gain can be expressed as follows [Lee et al. 2003];

$$\Delta V_I = \eta_{G,1}-\eta_{G,2} = q(u_1^{0.5}-u_2^{0.5}) \quad (3)$$

where η_G is the overvoltage due to the gas-phase mass transfer effect, q is the slope of the voltage gain vs. square root of utilization, u is the utilization, and subscripts 1 and 2 denote before and after N_2 addition, respectively. On the other hand, the voltage gain at the cathode ($\Delta V_{I,ca}$) is relatively small compared with $\Delta V_{I,an}$ as shown in Fig. 3(a). This means that the cathode has smaller mass-transfer resistance in the gas phase than anode because the cathode has only deficiency of O_2 species and more sluggish diffusion process through the liquid electrolyte film prevails [Lee et al. 2005].

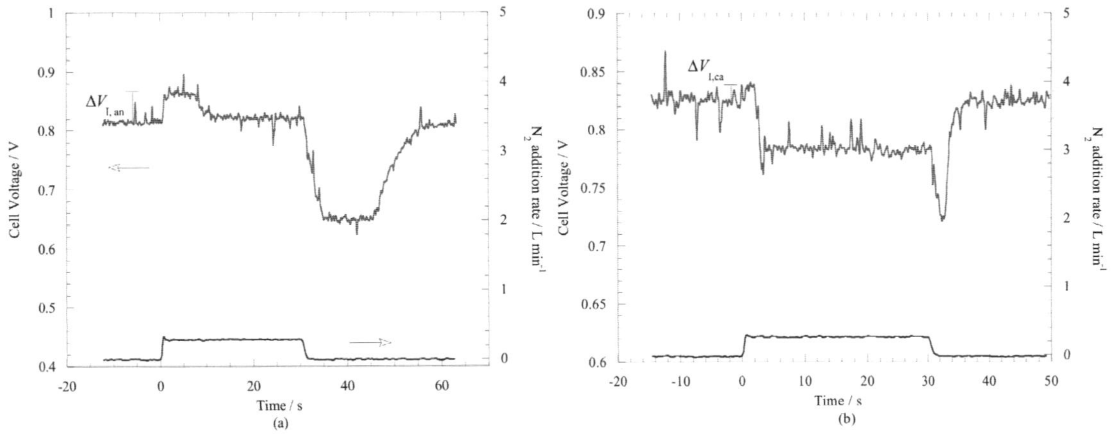

Fig. 3 Voltage patterns by the addition of 0.3 L min^{-1} N_2 to the anode of total flow rate of 0.253 L min^{-1} (a) and cathode of total flow rate of 0.883 L min^{-1} (b) at 923K, 1 atm.

When we assume that the utilization before the N_2 addition is a constant, then the following equation is available from Eq.(3) [Lee et al. 2003];

$$\Delta V_I = m-qu_2^{0.5} \quad (4)$$

Figure 4 shows the relation of Eq.(4). Since the anode gas flow rate (total 0.253 L min^{-1} at u_f=0.6) is much smaller than that of cathode (total 0.883 L min^{-1} at u_{ox}=0.4), the same amount of flow rate increase by the addition of N_2 (0.1 to 1.0 L min^{-1}) reduces the anode utilization to much lower value than the cathode utilization. We can obtain q values from the slopes; the anodic value of q (q_{an}) is 160.0 mV and the cathodic one (q_{ca}) is 29.5 mV. In addition, the overpotential due to the gas-phase transport (η_G) can be expressed as Eq.(5) from the Eq.(3).

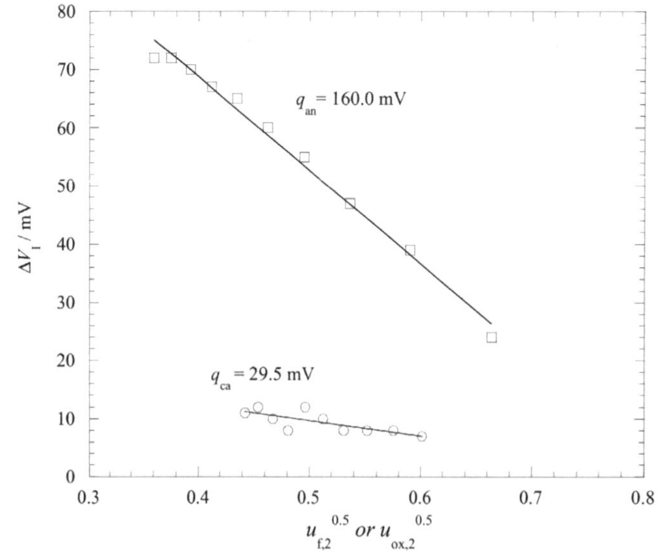

Fig. 4 Relations of Eq. (4) at the anode and cathode with ΔV_l obtained from Fig. 3.

$$\eta_G = qu^{0.5} \tag{5}$$

With Eq.(5) we can obtain the overpotential due to the gas-phase transport at various utilization.

The overpotential due to the gas-phase transport (η_G) was compared with the Nernst loss obtained by experimental way and calculation. As a experimental way for the Nernst loss, the constant gas utilization (CU) method was employed. An arithmetic mean of potential difference between the gas inlet and outlet (Nst) was calculated for the Nernst loss as a simple calculation (SC) method. Since the open circuit voltage is a function of temperature and gas compositions (Eq.(1)), the different gas compositions at the gas inlet and outlet at a current load gives potential gap.

$$Nst = (E_{in} - E_{out})/2 \tag{6}$$

Figure 5 compares the overpotential at the anode ($\eta_{an,G}$) and the cathode ($\eta_{ca,G}$) with Nernst loss with CU method ($\Delta V_{u,an}$ at the anode and $\Delta V_{u,ca}$ at the cathode) and SC method (Nst_{an} at the anode and Nst_{ca} at the cathode). Although the values at the anode have some of deviation, they show consistency of rising with utilization at a similar slope. The values among them show more close each other at the cathode. As mentioned above, the result of ISA indicates that MCFC has overpotential due to the slow gas-phase mass transfer process. This implies that the concentration gradient from the gas phase reactant to the electrode surface exists and the compositions of reactants in the bulk gas are different from those of electrode surface. However, the Nernst loss was suggested on the assumption of the equality of bulk gas composition to the electrode surface. When we consider the discrepancy between Nernst loss and the experimental results, the overpotential variation by utilization change is ascribed to the kinetic reasons of mass transfer in the gas phase.

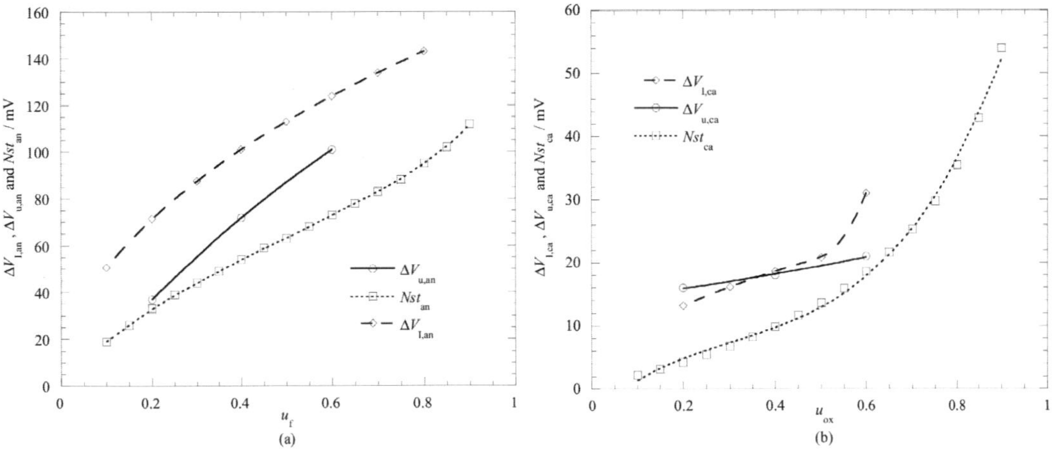

Fig. 5 Comparisons of overpotential measured by ISA method (ΔV_l) with Nernst loss obtained by CU method (ΔV_u) and calculation (*Nst*) at the anode (a) and cathode (b) at 923K, 1 atm.

4. Conclusion

The overpotential, showing larger value at higher utilization, was investigated by comparison of overpotential due to the gas-phase mass-transfer resistance with Nernst loss, a hypothetic voltage loss with thermodynamic backgrounds. The overpotential induced by the mass transfer effect was measured with inert gas step addition (ISA) method, whereas the Nernst loss was obtained by a calculation (simple calculation, SC) method and an experimental way (constant gas utilization, CU). Measured overpotential with ISA showed similar behavior against utilization with values of Nernst loss obtained by SC and CU methods. The Nernst loss, based on the assumption that the cell potential is determined by the bulk gas composition, is a kind of hypothetic value and discord with an experimental result that the bulk gas composition may be different from that of the electrode surface. When we consider the theoretical vulnerability of Nernst loss, the overvoltage is ascribed to the kinetic reason of mass transfer effect through the gas phase.

References

Lee C.-G., Kang B.-S., Seo H.-K., Lim H.-C. (2003). Effect of gas-phase transport in molten carbonate fuel cell, *J of Electroanal. Chemistry*, *540,* 169.

Lee C.-G., Oh J.-M., H.-C. Lim (2003). Electrode reaction characteristics with slight pressure change in a molten carbonate fuel cell, *J. of Electroanal. Chemistry*, *560*, 1.

Lee C.-G., Lim H.-C. (2005). Experimental Investigation of Electrode Reaction Characteristics with Reactant Gas Addition Measurement in a Molten Carbonate Fuel Cell, *J. of Electrochem. Soc.*, *152*, A219.

Sampath V., Sammells A. F., Selman J. R. (1980). A performance and current distribution model for scaled-up molten carbonate fuel cells, *J. Electrochem. Soc., 127*, 79.

Bard A. J., Faulkner L. R. (2004). Electrochemical Methods, *Wiley*.

Paper No. 690g

Estimation of Parameters for the Bohart-Adams Model

C.P. Thurgood

Electrochemical Power Sources Group
Dept. of Chemistry and Chemical Engineering
Royal Military College of Canada
Kingston, Ontario

Keywords: biogas, Fuel Cells, Bohart-Adams

Prepared for presentation at A.I.Ch.E. National Meeting, Philadelphia, PA
November 16 – November 21, 2007

Unpublished

Abstract

The vision of the Hydrogen Economy is to use renewable energy to power Fuel Cells. Power systems based on this pattern have reduced carbon footprint, minimal production of pollutants and potentially high efficiency. Biomass is of particular interest as large quantities of plant and wood (cellulose) residues are generated annually and are presently considered a waste product.

Anaerobic Digesters are a commercial, well established, technology for converting a wide range of biomass (manure and some plant residues) into a benign solids product and a methane rich gas (biogas).

A number of demonstration projects have shown that Fuel Cells can operate on biogas, a pattern of energy production that is a realization of the Hydrogen Economy.

The biogas contains a number of contaminants, such as hydrogen sulphide, that are detrimental to Fuel Cells. Currently, adsorbent beds are the primary technology for the removal of these containments. These beds must be sized correctly in order to protect the Fuel Cell but not oversized in order to minimize capital and operating costs.

The Bohart-Adams model is widely used to size fixed bed adsorbers. A critical role of the model is to fit it to breakthrough curves from laboratory beds, extract model parameters and use the parameters to design the commercial adsorber. With properly calibrated parameters, the design of the bed can proceed with confidence.

This article investigates the method of fitting the Bohart-Adams model to breakthrough data, using methanol adsorption on activated carbon as a test case, and computes the consequent error in the model parameters.

1 Introduction

A key element in the development of the Hydrogen Economy is the linking of Fuel Cells to a renewable energy resource such wind power, solar or biomass. The overall power system has many benefits. As the raw fuel is renewable, the power system is carbon neutral: there is no net addition of carbon dioxide to the atmosphere. The Fuel Cell ensures reduced or even no emissions of common pollutants such as NO_x, SO_x and particulates. In addition, Fuel Cells are potentially efficiency with efficiencies greater than the prevalent energy conversion technology: the internal combustion engine. If the Fuel Cell operates at high temperature (SOFC or MCFC), there is a further advantage as the high quality heat can be converted to additional power.

There are a number of well noted weaknesses in the vision of the Hydrogen Economy. First, renewable energy from wind or solar has significant variation with the time of day and season. There must significant means to store energy when it is being produced and not needed, so that it is available for use when demanded by the user. The requirement for storage capacity greatly increases the cost of power generation with the consequence that 'green' power is much more expensive then power generated from combustion of fossil fuels.

The weakness with biomass as a renewable energy lies in processing it to a useful form suitable for a fuel cell. Biomass is a complex material that varies with the plant. Some technologies show promise of being able to handle a wide range of plant material such as enzyme technology or gasification. However, these technologies are pre-commercial or still under research.

Anaerobic Digestion (AD) is one well established, commercial, technology able to process a wide range of biomass. AD technology is very attractive for decomposing animal manure (cattle, pig or chicken), sewage at waste water treatment plants, or kitchen waste in urban areas. The AD converts the 'waste' into a methane rich biogas and solids which can be used as a soil amendment. The biogas can be converted to power with a 'genset' which offsets the costs of the AD installation.

Moreover, the hydraulic retention time of an AD can be quite long (many days). The AD can 'store' biomass for significant periods smoothing variation in the feed of raw biomass.

The AD technology provides a critical link between a renewable energy source and Fuel Cells. There are a number of demonstration projects that show this link is technically feasible and that the biogas is suitable fuel for Fuel Cells [1].

A common element in these demonstrations is the cleanup of the biogas. Biogas contains a large number of impurities, such as hydrogen sulphide, that are detrimental to the Fuel Cell. One technology used to remove impurities are adsorbents such as activated carbon [2]. The correct sizing of the beds is important in protecting the Fuel Cell and the overall economics of the process.

The Bohart-Adams model is widely used to size fixed bed adsorbers [3]. A critical role of the model is to fit it to breakthrough curves from laboratory beds, extract model parameters and use the parameters to design the commercial adsorber.

In this article, we compare linear and non-linear methods to fit the Bohart-Adams model to a sample breakthrough curve.

Time (mins)	Exit Conc. (mg/m^3)
13	0.13
14	0.26
15	0.53
16	1.18
17	2.49
18	5.52
19	10.64
20	19.83
21	37.29
22	61.19
23	96.25
24	150.61

Table 1 Breakthrough data
For AGC exposed to methanol

2 Experimental

A standard breakthrough experiment was performed using a proprietary activated carbon. A canister was loaded with 175 cm^3 of the carbon and dehumidified (RH<15%) by dried air until a constant mass was achieved. The bed was then exposed to dried air at 30 SLPM containing 2007 mg/m3 of methanol (Aldrich). The exhaust from the bed was monitored by an IR detector that was calibrated for methanol concentrations from 0 - 150 mg/m3 (115 ppm). The data from the experiment are presented in Table 1. Methanol was detected in the exit stream after 13 minutes and reached the limits of the detector calibration after 24 minutes. The run was extended beyond 24 minutes but the data is not considered reliable due to a possible saturation of the IR monitor.

3 Bohart-Adams Model

The Bohart-Adams model is well established in the literature [3]. The key assumption of the model is the 'rectangle' or step isotherm: the equilibrium capacity of the adsorbent is fixed at q_s (mg/cm^3) and independent of the concentration of the species in the vapour phase. Application of a molar balance over a differential cylinder of the material gives the following governing equations:

$$\frac{\partial c}{\partial t}+v\frac{\partial c}{\partial z}+\frac{(1-\varepsilon)}{\varepsilon}\frac{\partial \overline{q}}{\partial t}=0 \qquad \text{Equation 1}$$

$$\frac{\partial \overline{q}}{\partial t}=kc(q_s-\overline{q}) \qquad \text{Equation 2}$$

In equation (1) and (2), c is the interstitial concentration of adsorbate (methanol, mg /cm^3 of void space), v the interstitial velocity (cm/sec), ε the void fraction, $\overline{q}$ the uptake of adsorbate (mg/cm^3 of adsorbent) and q_s the equilibrium uptake. The parameter k, the adsorption coefficient, has units of cm^3 of void space/ (mg-sec).

The bed is assumed to be homogenous and gradients occur only in the axial direction.

At the time t = 0, the bed is free of adsorbate ($\overline{q}=0$) and exposed to an inlet stream of inert gas carrying the adsorbate at a constant concentration of c_0. The solution to equations (1) & (2) is:

$$\frac{c}{c_0}=\frac{e^{\tau}}{e^{\tau}+e^{\xi}-1} \qquad \text{Equation 3}$$

where $\tau=kc_0\left(t-\frac{z}{v}\right)$ and $\xi=\frac{kq_s z}{v}\left(\frac{1-\varepsilon}{\varepsilon}\right)$. The transformed variables τ and ξ are, respectively, a dimensionless time and distance. The exit concentration is obtained by setting $z=L$, where L is the length of the bed.

The use of the Bohart-Adams model to design adsorbent beds requires accurate knowledge of the two parameters k and q_s. These parameters are not easily predicted since they represent the aggregate effects of mass transfer, pore diffusion, kinetics of adsorption and equilibrium.

It is standard practice to measure a breakthrough curve for the contaminated stream and a proposed adsorbent with flows, concentrations and compositions typical of the actual application. Equation (3) is fit to the breakthrough data to determine the two parameters. This calibrates the Bohart-Adams model so that it can be applied with confidence to size the bed.

It is custom to rewrite Equation (3) to facilitate the fit to the breakthrough data. The reciprocal of Equation (3), after some trivial algebra, can be written as:

$$\frac{c_0}{c}-1=e^{\xi-\tau}-e^{-\tau} \qquad \text{Equation 4}$$

The second term on the right side is usually small and can be neglected. We also recognize that for long times, the term z/v in the expression for τ can be ignored. Taking natural logarithms yields the equation:

$$\ln\left(\frac{c_0}{c}-1\right)=\frac{kq_s z}{v}\left(\frac{1-\varepsilon}{\varepsilon}\right)-kc_0 t \qquad \text{Equation 5}$$

A plot of LHS versus time yields a straight line. The tools of linear regression are used to compute the intercept and slope, from which it is simple to find k and q_s.

The alternative is non-linear regression: fit equation (3) directly to the data and determine the value of k and q_s that minimize the sum of squares.

It is worthwhile noting that equation (5) has the same form as the Wheeler-Jonas equation [4] which has broad use in the literature for calculating breakthrough times. Linear regression is also used to fits the model to data in order to determine the parameters. This investigation offers insight into the quality of the parameters for the Wheeler-Jonas equation.

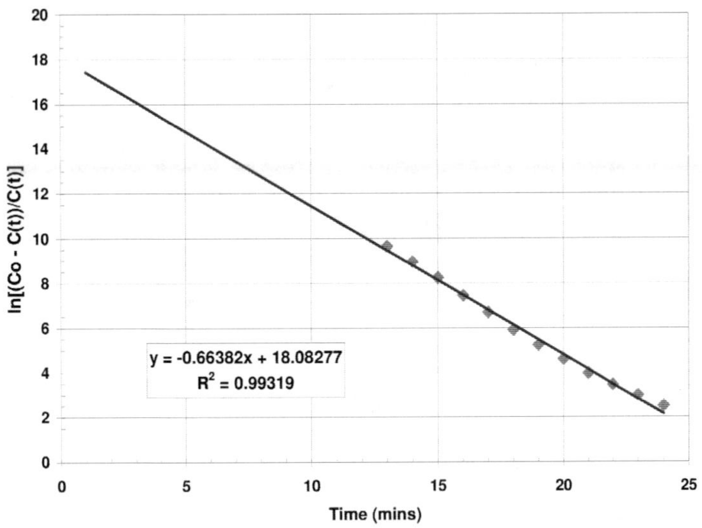

Figure 1 Linear Regression of methanol data according to Equation (5)

3.1 Linear Regression

The results of fitting Equation (5) to the data of Table 1 are presented in Figure 1. The methanol breakthrough data are well represented by Equation (5) as the R^2 = 0.9934 – most of the variance in the data is removed by the model. The values for k and q_s are, respectively, 5.5125 $cm^3 \cdot mg^{-1} \cdot sec^{-1}$ and 18.7446 $mg \cdot cm^{-3}$

The interdependence of slope and intercept (and also k and q_s) is evident in the graph. The intercept is obtained by extrapolating the regression backwards to zero time (as suggested by the backwards extrapolation of the trend line). Small changes in the slope lead to significant changes in the intercept. We also note that the transformation (LHS of Equation (5)) is not well defined as time approaches zero. The concentration at the exit of the bed is zero at t=0 leading to an undefined calculation. At times near zero, small measurement errors in the concentration are amplified via the transformation leading to large uncertainties in the intercept.

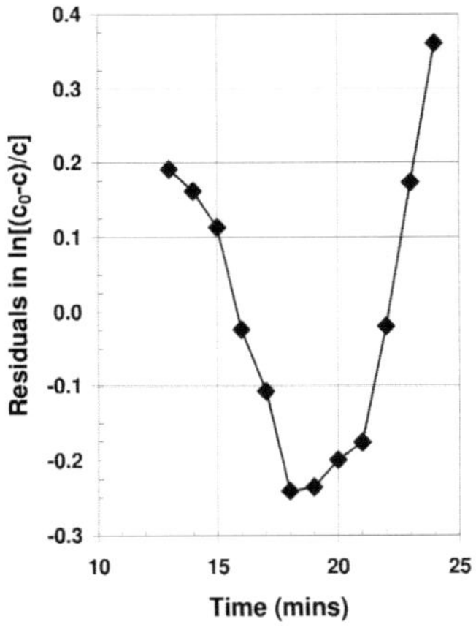

Figure 2 Residuals in linear regression of Figure 1. Residuals are evenly distributed about zero although there is a systematic variation indictaing a second order (parabolic) effect is present

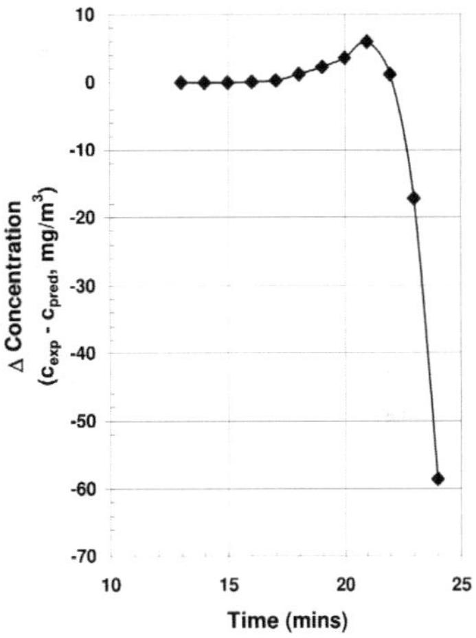

Figure 3 Residuals in concentration derived from linear regression of Figure 1. Residuals are not evenly distributed about zero and there is a systematic variation with time

The residuals of the fit for equation (5) are presented in Figure 2. The residuals show that there is a small systematic trend in the data – a second order term could be included in Equation (5) to improve the fit. However, they are evenly distributed about zero and the high R^2 of 0.9913 indicates that equation (5) is an excellent representation of the transformed data.

Figure 3 presents the residuals in the concentration. This is the difference between the data and the concentration computed using Equation (3) with the values for k and q_s computed from the linear regression. It is quite clear that there is a bias. The residuals increase significantly with time rather being evenly distributed about zero as in Figure 2. The predicted values for the concentration are presented in Table 2.

An exposition of the theory and calculation of the joint confidence region for the parameters in a linear model is given in [5]. The general linear model is written as:

$$Y = X\beta + Z \qquad \text{Equation 6}$$

where Y is the response (concentration), X the regressor variable (time) and Z, a random variable representing error in the model. The variable β are the parameters, the slope and intercept of equation (5).The fit of the model to data is obtained by minimizing the sum of squares:

$$S(\beta) = \|y - Y\|^2 \qquad \text{Equation 7}$$

Carrying out the least squares fitting procedure yields the best estimates of the parameters, called $\hat{\beta}$. The estimates can be calculated by the expression:

$$\hat{\beta} = \left(X^T X\right)^{-1} X^T Y \qquad \text{Equation 8}$$

The joint confidence region for the parameters β at the α confidence level is the ellipsoid:

$$\left(\beta - \hat{\beta}\right)^T X^T X\left(\beta - \hat{\beta}\right) \le P \cdot s^2 \cdot F(P, N-P; 1-\alpha) \qquad \text{Equation 9}$$

where P is the number of parameters in the model, N is the number of data points and F(P,N-P;1-α) is Fishers F distribution with P and N-P degrees of freedom. The residual mean square, s^2, is calculated from the expression:

$$s^2 = \frac{S\left(\hat{\beta}\right)}{N - P} \qquad \text{Equation 10}$$

For the data in Table 1, X^T=[1 13; 1 14; 1 15; ...,1 23; 1 24] and

$$X^T X = \begin{bmatrix} 12 & 222 \\ 222 & 4250 \end{bmatrix} \qquad \text{Equation 11}$$

Other values entering into equation (9) are: P = 2; N-P = 10; F(2,10;0.05) = 4.10 and s^2 = 0.04323 ($S\left(\hat{\beta}\right)$ = 0.4328). The RHS of equation (9) is 0.3545.

Equation (9) develops into a quadratic that can be solved for combinations of slope and intercept of Equation (3) that defines the boundary of the 95% confidence region centred on $\hat{\beta}$. From the slope and intercept, the 95% confidence region for k and q_s can be developed.

The results of the calculations are presented in Figure 4. The confidence region is a symmetric ellipsoid while the range of k and q_s is surprisingly large. The adsorption coefficient varies from about 5.1 to 5.9 around the optimum value is 5.5. This corresponds to a relative variation of about 7% (0.4/5.5*100). The capacity varies from 18.3 to 19.3 around the optimal value of 18.7 or a relative variation of about 3% (0.6/18.7*100).

3.2 Non-Linear Regression

The non-linear regression involves minimizing the sum of squares (equation (7)) where the model predictions (Y) are provided by equation (3). The initial guess for the parameters k and q_s was the results of the linear regression. Indeed, the numerical routine (SOLVER function in Excel) had great difficulty locating the minimum sum of squares without a good initial estimate. The minimum was checked by preparing a two dimensional matrix of values k and q_s, computing the sum of

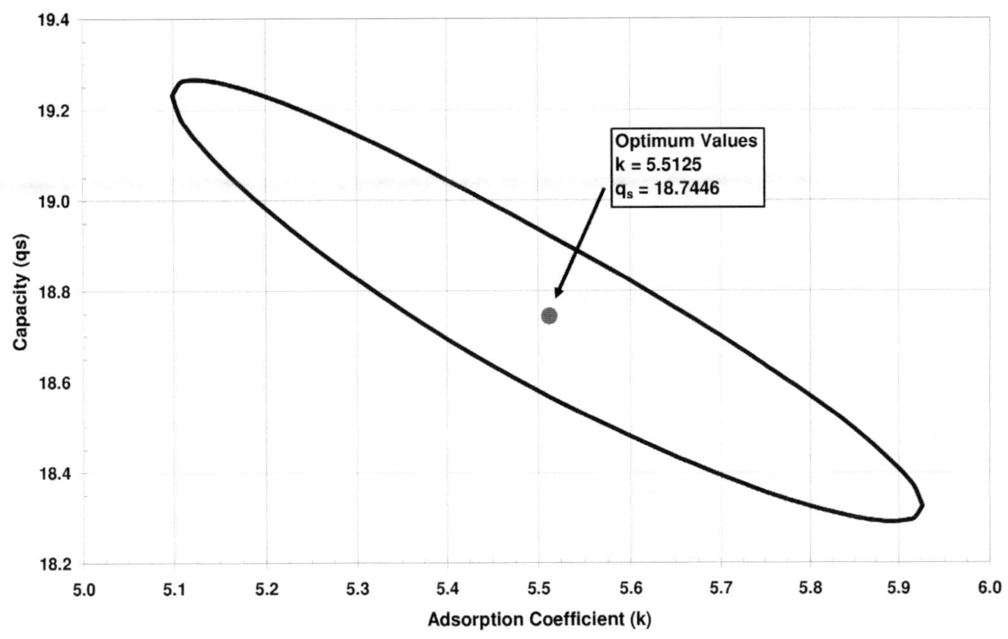

Figure 4 The joint confidence region (95% confidence level) for k and q_s based upon linear regression. The blue line is the boundary of the confidence region and the red dot is the value of k and q_s that minimize the sum of squares (Optimum Values).

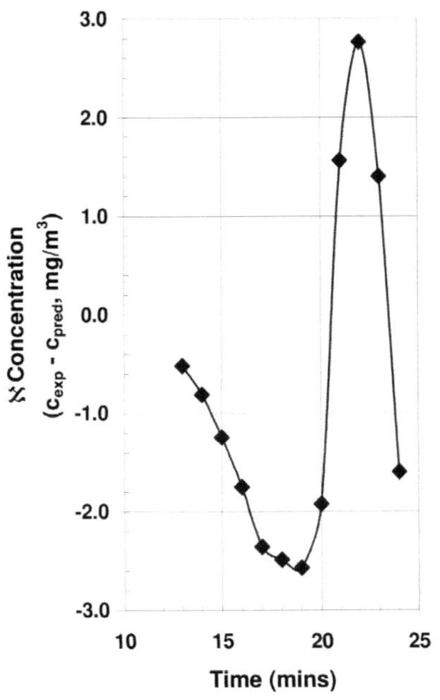

Figure 5 Residuals in concentration derived from the non-linear regression. Residuals are evenly distributed about zero but show a sinusoidal (systematic) variation with time

squares for each combination and manually checking for the minimum sum of squares. These efforts lead to values for k and q_s of, respectively, 4.1804 cm^3·mg^{-1}·sec^{-1} and 19.9305 mg·cm^{-3}.

The residuals in the concentration of the fit of equation (3) are presented in Figure 5. The residuals are evenly distributed about zero but show a sinusoidal variation that suggests a small systematic trend. This is consistent with the similar observation for the linear regression.

The predicted values for the concentration for the linear and non-linear fit are presented in Table 2. The non-linear fit gives much better representation of the data at larger times, especially for times 21 – 24 minutes. The relative error at small times (13 – 15 minutes) is much larger than the error in the linear fit but is still close to the data.

The calculation of the confidence region for the non-linear estimates of the parameters is given by the equation [5]

$$S(\beta) \le S(\hat{\beta})\left(1 + \frac{P}{N-P} \cdot F(P, N-P; 1-\alpha)\right) \quad \text{Equation 12}$$

Values entering into equation (12) are: P = 2; N-P = 10; F(2,10;0.05) = 4.10, $S(\hat{\beta}) = 42.0925$. Carrying out the calculations gives $S(\beta) \le 76.8213$.

A trial and error method was used to determine values of β that satisfied equation (12) exactly (i.e. $S(\beta) = 76.8213$). This procedure was greatly assisted by the matrix of values, mentioned above, to confirm the location of the minimum sum of squares.

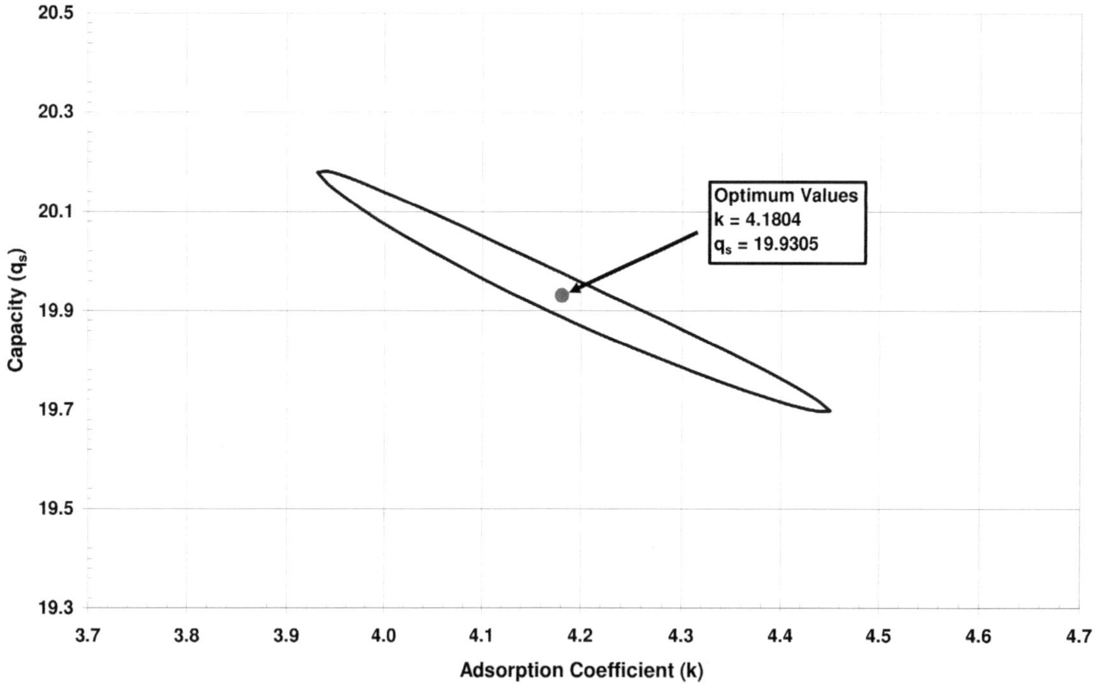

Figure 6 The joint confidence region (95% confidence level) for k and q_s based upon non-linear regression. The blue line is the boundary of the confidence region and the red dot is the value of k and q_s that minimize the sum of squares (Optimum Values). The scales of the graph are the same as Figure 4 to facilitate comparison.

Time	Exit Concentration (mg/m^3)		
(mins)	Measured	Linear Fit	Non-linear Fit
13	0.13	0.16	0.65
14	0.26	0.31	1.07
15	0.53	0.59	1.77
16	1.18	1.15	2.93
17	2.49	2.24	4.84
18	5.52	4.34	8.00
19	10.64	8.41	13.20
20	19.83	16.28	21.75
21	37.29	31.38	35.73
22	61.19	60.06	58.43
23	96.25	113.44	94.85
24	150.61	209.19	152.20

Table 2 Exit concentrations, computed from the linear and non-linear model fits, compared to the experimental data. The linear fit is superior for low concentrations while the non-linear fit is superior for the high concentrations.

The results of the calculations are presented in Figure 6. The confidence region is an apparently symmetric ellipsoid. Increasing the confidence level shows that the joint confidence region is asymmetric and is banana shaped. The adsorption coefficient varies from about 3.93 to 4.45 around the centre value of 4.18. This corresponds to a relative variation of about 6% (0.26/4.18*100). The capacity varies from 19.7 to 20.2 around the optimal value of 19.93 or a relative variation of about 1.3% (0.25/19.93*100).

The two regression techniques produce best estimates that are significantly different at the 95% confidence level. That is, the two confidence regions do not intersect and there is no combination of k and q_s that will generate an error that meets the criteria of both equation (9) and (12) simultaneously.

Inspection of the two figures clearly shows that the area of the non-linear confidence region (Figure 6) is smaller than the area of the linear confidence region (Figure 4). The area of the confidence region is estimated by using the trapezoidal rule. Figure 4 has an area of 0.2315 while the area in

Figure 6 is 0.01767, a factor of 10 smaller. The area reflects the random error of the underlying measurements. The fact that Figure 4 has greater area than Figure 6 indicates that fitting procedure (linear regression) is inflating the error. For this data, non-linear regression generates tighter bounds on the parameter estimates

Another metric of the joint confidence region is its 'thickness'. Consider Figure 4 and a fixed value for the adsorption coefficient. The values of the capacity that lie on the vertical line and within the joint confidence region provide satisfactory fits to the data at the 95% confidence level. Clearly, the wider the region at the given value of the adsorption coefficient, the larger is the uncertainty in the true value of the capacity. The same analysis can be carried out for a horizontal line in this case the thickness is a measure of the uncertainty in the adsorption coefficient for the given capacity.

In view of this, Figure 6 is preferred over Figure 4 since it is slender ellipsoid and its width in both vertical and horizontal directions is smaller.

Now, it is desired that the ellipsoid collapse to a vertical or horizontal line segment. In either case, the estimates of the parameters are independent of each other. For a horizontal line – the uncertainty in the adsorption coefficient has no effect on the capacity; for a vertical line segment the uncertainty in the capacity has no effect on the adsorption coefficient.

In view of this, a final metric of the confidence region is its 'tilt'. This refers to the angle the region makes with respect to the vertical and is a gross measure of the degree of dependence of the two parameters. The dependence is reflected that if one parameter undergoes a change, a proportionate change in the other parameter is required in order to maintain the same sum of squares or goodness of fit. For a vertical or horizontal line segment, the dependence is zero or absent, and it is possible to estimate the parameters independently.

The extremities of the confidence region are used to define a straight line that lies close to the optimum values and approximately bisects the region in two. For Figure 4, the slope of the line is -1.0972 or an angle of 42° from the horizontal. For Figure 6, the slope of the line is -0.9094 or an angle 48° from the horizontal. Both regions indicate a similar dependence between the parameters.

4 Summary

The Bohart-Adams can be fit to measured breakthrough data in two different ways. The conventional method involves linearization of the model (see equation (5)). The fitting can be conveniently done in a commercial software (excel) with automatic generation of relevant statistical metrics and tests. The non-linear fitting is more involved and finding the parameters that correspond to the true minimum in the sum of squares is difficult. Moreover, the construction of the joint confidence region, the key entity in assessing the quality of the fit and the error in the parameters, is non-trivial.

The two methods generate statistically different parameter estimates but these estimates differ by much less than a factor of 2. For the data presented in this article, the difference in the estimates is on the order of 30% for the adsorption coefficient and 6% for the capacity.

The advantage of the non-linear fit procedure is that the joint confidence region seems to be significantly smaller than the corresponding region generated by linear regression. This leads to better estimates of the parameters (smaller error). However the degree of correlation between the parameters is quite similar for the two fitting methods.

These conclusions are strictly limited to the experimental data presented in this article. However, these findings motivate carrying out a similar investigation on a broader collection of data to confirm advantages of non-linear over linear regression.

5 Acknowledgements

This work is supported by the Canadian Department of National Defence through the Defence Research and Development Branch and by a National Science Engineering and Research Discovery Grant.

6 References

1. http://www.fuelcells.org/basics/faqs.html accessed September 20, 2008 at 11:09 am
2. A. Bagreev, S. Katikaneni, S. Parab and T.J. Bandosz, Catalysis Today 99 (2005) 329 - 337
3. M.D. Ruthven, Principles of Adsorption and Adsorption Processes, Wiley, NY, 1984
4. P. Lodewyckx, G.O. Wood, S.K. Ryu, Carbon 42 (2004) 1351 - 1355
5. D.M. Watts and D.G. Bates, Nonlinear Regression Analysis and Its applications, Wiley, NY, 1988

708a Analysis of Palladium Based Anode Electrode Using Electrochemical Impedance Spectra In Direct Formic Acid Fuel Cells

Won Suk Jung[1], Jonghee Han[1], and **Su Ha**[2]. (1) Fuel Cell Research Center, Korea Institute of Science and Technology, 39-1 Hawolgok-dong, Seongbuk-gu, 136-791, Seoul, South Korea, (2) School of Chemical Engineering and Bioengineering, Washington State University, PO Box 642710, Pullman, WA 99164-2710

In this study, we used the electrochemical impedance spectra to evaluate the anode performance of direct formic acid fuel cell (DFAFC), and how its anode charge transfer resistance ($R_{anode,ct}$) and electrolyte resistance (R_{ele}) are affected by various cell operating parameters. These parameters that we have investigated in this study include the anode overpotentials, cell operation times, formic acid feed concentrations and cell temperatures. The anode impedance spectra demonstrated that the $R_{anode,ct}$ and R_{ele} are low for the DFAFC using 5M formic acid feed concentration, which leads to its high power density output of 250 mW/cm^2 at 0.35 V and 30°C. This high performance of the DFAFC shows that it has a great potential for portable power applications. The $R_{anode,ct}$ increases gradually as either the cell operation time increases or the formic acid feed concentration is raised from 10M to 15M, which leads to deactivations of the anode and the overall cell performances. However, these deactivation processes are reversible and the cell performance can be easily reactivated.

708b Improved Operation of the Direct Formic Acid Fuel Cell by the Use of Electrochemically Modified Anode Catalysts

John L. Haan, Department of Chemistry, University of Illinois at Urbana-Champaign, 600 S Mathews, Urbana, IL 61801 and Richard I. Masel, Chemical and Biomolecular Engineering, University of Illinois at Urbana-Champaign, 600 S. Mathews Ave, RAL 204, Urbana, IL 61801.

The direct formic acid fuel cell (DFAFC) was developed as an alternative energy source for portable electronic devices. The DFAFC uses a safer fuel that exhibits lower crossover than methanol. In the DFAFC, one area of losses is at the anode, where kinetics can be improved and poisoning reduced. It is possible to decrease these losses through the use of electrochemically modified catalysts. Performance of the palladium black anode catalyst is improved when adatoms such as antimony, lead, tin and vanadium are electrochemically added to the palladium surface without significantly altering the bulk catalyst composition. Improvements have been observed both in an operating fuel cell as well as a three electrode electrochemical cell.

708c A Dual Porosity Carbon Support for Direct Methanol Fuel Cells

Kwong-Yu Chan[1], Siu Wa Ting[2], and Jiawen Ren[2]. (1) Department of Chemistry, The University of Hong Kong, Pokfulam Road, Hong Kong, Hong Kong, (2) Chemistry, University of Hong Kong, Pokfulam Road, Hong Kong, Hong Kong

Carbons with well defined porous structures are synthesized using different types of mesoporous silcas as templates. The silica templates include SBA-15, SBA-16, FDU-12, and M41S, leading to carbon structures with different pore size and distribution corresponding to the templates. These carbons are used as supports for platinum-ruthenium nanoparticles and tested for activity towards methanol oxidation in a fuel cell electrode. The mixed metal nanoparticles are synthesized by the ethylene glycol method with good control of size and composition. The carbon structures as well as the silica parents were characterized by

transmission electron microscopy (TEM), X-ray diffraction pattern (XRD) and nitrogen sorption analyses. Performance of the electrocatalysts is determined by polarization curves and AC impedance measurement. Attempts are made to correlate catalyst performance to the structural parameters of the carbon including pore size, pore distribution, surface area, and pore volume. It is shown that a single structural parameter such as surface area or pore volume cannot determine performance. The arrangement of the porous network at different length scales collectively determine performance especially at higher current densities when ionic and mass transport become important. A dual porosity structure synthesized with M41S silica of controlled size shows the best performance. This carbon has a meso-cellular foam like carbon structure composed of 30 to 50 nm porous cells with internal hexagonally arranged channels of 3 nm diameter. The smaller pores favor anchoring of PtRu nanoparticles whereas the larger channels of 13 nm between cells favor transport. The dual porosity network structure had a high surface area (> 1000m2/g) and large pore volume (> 1.2 mL/g). When loaded with Pt-Ru nanoparticles, the PtRu/carbon became a good electrocatalyst and gave higher activity for methanol oxidation compared to a commercial catalyst.

Acknowledgment

Research reported in this paper has been supported by the Research Grants Council of Hong Kong (HKU and the Sino-German Collaboration Research. This work has been supported by the Research Grants Council of HongKong (HKU7072/01P,HKU7005/03P), a Joint Sino-German Collaborative Research Grant on "Fundamental Aspects of Direct-Alcohol Fuel Cells", and a seed grant from the HKU Foundation.

References

[1] Chan, K.-Y., Ding, J., Ren J., Cheng S., and Tsang, K. Y., J. Mater. Chem., 14 (2004) 505.

[2] Ding J, Chan KY, Ren J, Xiao F., Electrochimica acta 2005; 50: 3131-3141.

[3] Ren, J., Ding, J., Chan, K.Y., Wang, H.T., Chem. Mater. 19 (2007) 2786.

708d A Comparative Study of Tungsten Monocarbide and Platinum as Counter Electrode Materials In Photoelectrochemical Solar Cells

Daniel V. Esposito[1], Kevin D. Dobson[2], Brian E. McCandless[2], Robert W. Birkmire[2], and Jingguang G. Chen[1]. (1) Chemical Engineering, University of Delaware, 150 Academy St., Newark, DE 19716, (2) Institute of Energy Conversion, University of Delaware, 451 Wyoming Rd., Newark, DE 19716

Despite their advantages over traditional solid-state solar cells, photoelectrochemical solar cells (PECs) have yet to enter the market place due to the absence of electrode materials that are efficient, low-cost, and stable in aqueous environments [1]. In the majority of PEC systems Platinum is used as a stable counter electrode material exhibiting low overpotentials, but its high cost and limited supply make it impractical for large-scale production of PECs. In this work, we explore the use of alternatives to replace Pt, including transition metal carbides and bimetallic alloys, as counter electrode materials in PECs. Numerous transition metal carbide [2] and bimetallic [3] materials are known to exhibit catalytic activity comparable to that of Pt-group metals in a variety of reaction systems, making them promising candidates for the replacement of Pt in those systems.

Research completed to date has focused on the use of tungsten monocarbide (WC) as a counter electrode material in an aqueous polysulfide PEC system. PECs based on polysulfide electrolytes and II-VI semiconductors have attracted much attention owing to their potential as low-cost photovoltaic systems [4]. Characterization of the stability and activity of WC and Pt electrodes has been performed using combined surface science (X-ray photoelectron spectroscopy) and electroanalytical (linear sweep voltammetry, chronoamperametry, cyclic voltammetry) methods. Both WC and Pt electrodes were seen to be stable at all potentials cathodic to their rest potential, but WC displayed a significant advantage in terms of activity relative to Pt in the polysulfide electrolyte. In this work we also explore the use of WC electrodes modified with submonolayer Pt. In previous work, our group has shown that the deposition of submonolayer Pt onto WC can have synergistic effects on the electrochemical stability and activity of the electrode [5].

An integrated PEC device comprised of a polysulfide electrolyte, thin-film CdSe photoelectrode, and WC or Pt counterelectrode has been constructed. The device was tested under AM 1.5 illumination for two hours, with no degradation in the short-circuit current being observed during this time. The PEC employing WC as the counterelectrode had a larger short-circuit current that that utilizing Pt, a result which is consistent with the electroanalytical experiments performed on the individual counter electrodes.

[1] T. Bak, J. Nowotny, and C.C. Sorrell, Int. J. of Hydrogen Energy, 27 (2002) 994-995.

[2] H. Hwu and J.G. Chen, Chemical Reviews, 105 (2005).

[3] J.G. Chen, C.A. Menning, and M.B. Zellner, Surface Science Reports, 63 (2008).

[4] J. Winnick, P.M. Lessner, and E.J. Cairns. Appl. Phys. Lett., 53 (1988) 1985.

[5] E.C. Weigert, A.L. Stottlemyer, M.B Zellner, and J.G. Chen, J. Phys. Chem. C, 111 (2007) 14617.

[4] J. Winnick, P.M. Lessner, and E.J. Cairns. Appl. Phys. Lett., 53 (1988) 1985.

[5] E.C. Weigert, A.L. Stottlemyer, M.B Zellner, and J.G. Chen, J. Phys. Chem. C, 111 (2007) 14617.

708e Synthesis and Electrocatalytic Property of Platinum-Based Heteronanostructures

Zhenmeng Peng and Hong Yang. Chemical Engineering, University of Rochester, 206 Gavett Hall, Rochester, NY 14627

Heterogeneous nanostructures that contain discrete domains of platinum and other metals are attractive candidates as advanced fuel cell catalysts that integrate several different functionalities. In this talk we present the synthesis of platinum-based heterogeneous nanostructures in various non-hydrolytic systems. We show the controls over the nucleation and growth of nanoparticles are possible. Platinum nanoparticles can be deposited on different supports through the reduction of platinum salt precursors at elevated temperatures. Bifunctional long chain surfactants were used as capping and reducing reagents. The electrochemical properties of these nanoparticles were characterized in solutions. These catalysts showed enhanced performances in oxygen reduction reaction (ORR), methanol oxidation reaction (MOR) and formic acid oxidation reaction (FAOR) in comparison with pure Pt catalysts.

708f The Electrochemical Impedance Study and Performance of Methanol Electrode Oxidation at Electrocatalyst Pt-Ru/C and Pt-Ru/CNT

Yu-Nong Chen[1], Ning-Yih Hsu[2], and **Shi-Chern Yen**[1]. (1) Chemical Engineering Department, National Taiwan University, Department of Chemical Engineering, National Taiwan University, Taipei, 10617, Taiwan, (2) Institute of Nuclear Energy Research (INER), P.O. Box 3-19, 1000 Wenhua Road, Longtan, Taoyuan 32546, Taiwan, Taoyuan, Taiwan

Electrochemical impedance was used to analyze the reaction kinetics and interfacial characteristics of an anode in a direct methanol fuel cell (DMFC). An advanced equivalent-circuit model is proposed. The new model incorporates constant phase elements (CPEs) rather than conventional capacitors in the equivalent-circuits taking into account the porous structure of the anode, particularly that in the catalyst layer and at the anode/membrane interface. It effectively simulated the electrochemical characteristics of a DMFC porous anode. The impedance model incorporates the interface factor, resulting in excellent matches between the simulation results and the experimental data in the Nyquist and the Bode plots over a wide range of frequencies. The performance of a single-cell direct methanol fuel cell (DMFC) using carbon nanotube-supported Pt-Ru (Pt-Ru/CNT) as an anode catalyst has also been investigated. In this study, the Pt-Ru/CNT electrocatalyst was synthesized using a modified polyol approach with a controlled composition very close to 20wt%Pt-10wt%Ru, and the anode was prepared by coating Pt-Ru/CNT electrocatalyst on a wet-proof carbon cloth substrate with a metal loading of about 4 mg cm-2. The membrane electrode assembly (MEA) was fabricated using Nafion 117 and the single-cell DMFC was assembled with graphite endplates as current collectors. Experiments were carried out at moderate low temperatures using 1 M CH_3OH aqueous solution and pure oxygen as reactants. The tested cell significantly outperformed a comparison cell using a commercial anode coated with carbon-supported Pt-Ru (Pt-Ru/C) electrocatalyst of similar composition and loading. High conductivity of carbon nanotube, good catalyst morphology and suitable catalyst composition of the prepared Pt-Ru/CNT electrocatalyst are considered to be some of the key factors for enhancing cell performance.

Acknowledgement

The authors would like to thank the National Science Council of Taiwan for financially supporting this research.

Catalytic Activity and Stability of Tungsten Oxide Electrocatalyst for Fuel Cell Applications

Shamalee A. Whitelocke and Egwu E. Kalu
Department of Chemical & Biomedical Engineering
FAMU-FSU College of Engineering
Tallahassee, FL

Introduction

The degradation of the electrocatalyst in the polymer electrolyte membrane fuel cell (PEMFC) is a major failure mode by which the performance of PEMFC deteriorates. Several factors - acting independently or in concert with each other - give rise to catalyst degradation. These factors include catalyst particle ripening (sintering), catalyst dissolution and catalyst support corrosion. Thus, different degradation mechanisms can occur on the electrode/electrocatalyst layer. The support physically separates the catalyst nanoparticles and may also influence the adsorptive properties of the catalyst active center. High electronic conductivity is a key performance requirement for any electrocatalyst support but unfortunately it also forms the basis of corrosion degradation. The presence of water in the catalyst/electrode layer promotes the corrosion of fuel cell components. Thus, carbon or graphitic carbon being the most commonly used support material in fuel cells is subject to corrosion degradation at very low potentials:

$$C + 2\mathrm{H}_2O \rightarrow CO_2 + 4\mathrm{H}^+ + 4e \qquad E_0 = 0.207\,V \ \ vs. \ \ RHE$$

However, because the rate of carbon corrosion at 0.207 V is slow, it can still be used as catalyst support. In fact Jarvi *et al.* [1] showed that for Vulan XC72, a substantial surface oxidation occurs at 65°C when the potential is cycled beyond 0.8 V vs. RHE. Platinum is the most active catalyst for fuel cells. However, the Pt deposited on the carbon substrate is known to catalyze the corrosion or oxidation of the carbon substrate with the formation of carbon dioxide as well as other active surface species such as quinone group [2]. Understanding the degradation mechanism is pertinent to proferring a viable solution. Efforts abound in the literature in understanding the underlying issues and evaluation of alternative catalyst supports that may be resistant to corrosion [3 – 6]. Some of the alternative materials to carbon that have been studied to date include the oxides and carbides of titanium and tungsten [3 – 6]. Because of the low conductivity of the oxides and carbides, efforts have been put in examining optimal combination of carbon (for improved conductivity) and oxides (or carbides) (for improved corrosion) that will result in durable and stable fuel cell.

For the oxide supports examined in the literature, Pt deposited through the impregnation method has often been the active catalyst most discussed. The performance degradation of Pt deposited by classical impregnation method has therefore been thoroughly examined by Gasteiger *et al.* [7] and reviewed in [8 – 9]. In the present work, cyclic voltammetric method is used to examine the catalytic activity of tungsten oxide support catalyzed by electroless method for oxygen reduction. Both Co-Ni and Pd catalysts are examined for their oxygen reduction performance as catalysts. Catalytic activity of C/WO_3 mixture for oxygen reduction is also examined.

Experimental Method

In preparing electroless Co-Ni catalyzed WO_3 (Co-Ni/WO_3) or (Co-Ni/WO_3-C), we started with polymer-stabilized Pd nanoparticles for the activation of the substrate for electroless deposition. The polymer-stabilized palladium catalyst solution was prepared by dissolving 0.2 g analytical reagent grade palladium acetate [$CH_3CO_2)_2Pd$] (47.05 % Pd) from Aldrich-Sigma Co, in 2.0 ml NH_4OH. The palladium mixture was then added with stirring to a 44 g methanol-poly(vinyl) butyral (Solutia Inc.'s Butvar B-98) mixture formed by dissolving 22 g PVB in 140 ml methanol. This is referred to as the Pd-ink. A known amount of tungsten oxide (Aldrich, St. Louis, MO)-carbon (Vulcan X72) was mixed with known amount of the Pd-ink and sonicated for 15 minutes. The Pd-ink/WO_3-C mixture was dried at 50°C in an oven and then subjected to 300°C temperature in a furnace for 24 hours. The different WO_3/C ratios investigated are: 100%/0%, 80%/20%, 50%/50% and 0%/100%.

A Co-Ni electroless solution for the deposition of electrocatalyst on the catalyzed substrate was prepared with analytical grade chemicals and deionized water. The solution contained 2.1 g $CuSO_4$, 4.9 g $NiSO_4$, 28g Tartrate, 20g NaH_2PO_2, 10g Boric Acid and 15g Gluconic Acid in a 1000 mL solution. The electroless plating temperature was fixed between 55 – 65°C and the pH maintained between 9.3 and 9.6 pH units with ammonium hydroxide used for the adjustment. Electroless plating was carried out for 2 minutes, 5 minutes and 10 minutes with three repeats for each time condition. The plated sample was filtered and washed with DI water. After drying the sample in an oven at 50°C, it was annealed at 250°C for 15 minutes.

The catalyst ink for oxygen reduction was prepared by mixing the electroless Co-Ni catalyzed powder (100 mg), Nafion solution (Dupont, 5% solution), water and n-propanol. Typically, equal volumes of Nafion, water and propanol is mixed and sonicated prior to the addition of the catalyzed powder. The resulting ink mixture was sonicated for about 15 minutes and about 2 μL drop applied on 3mm graphite electrode. After drying at room temperature, the resulting electrode was tested in a three-electrode set-up for oxygen reduction in 0.5 M H_2SO_4 solution.

Results and Discussion

Electrochemical Characterization

Fig. 1 shows the cyclic voltammetry (CV) of Co-Ni catalyzed substrates in oxygen saturated 0.5 M sulfuric acid. The Co-Ni was electrolessly plated for 2 minutes in each case. At 0.8 V (vs. Ag/AgCl), the figure shows that the WO_3/C ratio of 80%/20% mixture was the most active for oxygen reduction. This is unexpected since the low performing WO_3 would have been expected to decrease the activity of carbon. The figure thus suggests that the limitations in the activity of WO_3 is probably related to its low electronic conductivity as compared to carbon. A further comparison is made in figure 2 for oxygen reduction using different catalyzed substrates at 0.5 V (vs. Ag/AgCl). Again, similar to our findings in Fig. 1, the 80% WO_3/20% C mixture was most active for the Co-Ni catalyzed samples. This was followed by the 50% mixture. Here again, the performance of the mixed substrates show improvement over the individual components. In fig. 2, we observe that for both WO_3 and C, the deposition of electroless Co-Ni on Pd-catalyzed substrate conferred an improved activity over the Pd-catalyzed substrate. In Fig. 3, the effect of tungsten oxide composition on the catalyst activity for the reduction of oxygen is compared at 3 different CV scan rates. The trend observed is the same for each scan rate. The results suggest that an optimized mixture of WO_3/C can be obtained for an improved activity.

The effect of annealing time on the catalytic activity of Pd-catalyzed substrate was evaluated for

WO_3. Fig. 4 compares the catalytic activity of Pd-catalyzed WO_3 annealed for different time lengths. The sample annealed for a longer time period performed better – evidence that some left-over polymer could be covering the Pd nanoparticles. Annealing temperature and time are critical parameters for the polymer-stabilized nanoparticles.

Microstructure Characterization

Fig. 5a and 5b show the surface morphology of a 50% WO_3/50% C substrate and a spectrum (EDAX) of a selected area on the substrate, respectively. The spectrum shows the presence of Co, Ni and P deposited from the electroless process. The X-ray diffraction (XRD) results of Co-Ni catalyzed substrates and Pd catalyzed WO_3 are shown in Fig. 6. The absence of any strong peaks for the carbon catalyzed sample is noted. The peaks associated with tungsten oxide at different phases are evident. The presence of Co-Ni solid solution is not pronounced because of the interactions with tungsten oxide peaks.

Catalyst Stability

For a simple stability test, the catalysts were subjected to a constant potential hold at 1.7 V (vs. Ag/AgCl) for 30 minutes and this was followed by a cyclic voltammetry. At 1.7 V, it is expected that the metal-nanoparticle catalyst will be oxidized and carrying out a simple CV will indicate the reduction of this oxide in the cathodic direction. Based on the results, one can relate the stability of the catalyst to the reduction current of the oxide formed. Our results are inconclusive since the stability test was carried out in the presence of oxygen.

Conclusion

Electroless deposition was used to catalyze tungsten oxide and carbon/tungsten oxide mixtures for electrocatalysis of oxygen. The preliminary results obtained showed that this approach is feasible in improving the catalytic activity of carbon. Further optimization and stability studies are required in order to evaluate the potential of the approach. The use of polymer-stabilized nanoparticles for the catalyzation of substrate for fuel cell applications is demonstrated.

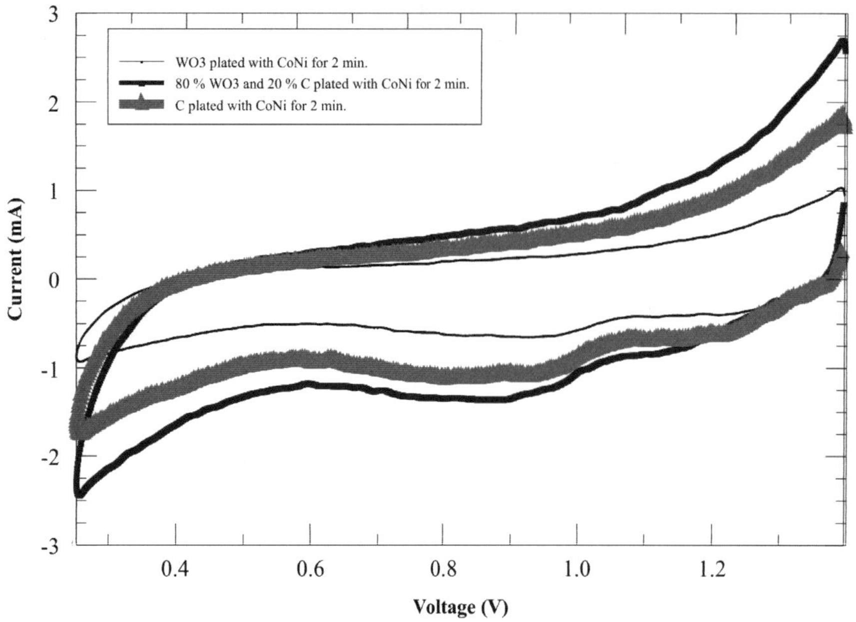

Figure 1. The Cyclic Volatmmetry graph of WO_3, 80% WO_3 and 20 % C, and Carbon all plated with CoNi for 2 minutes.

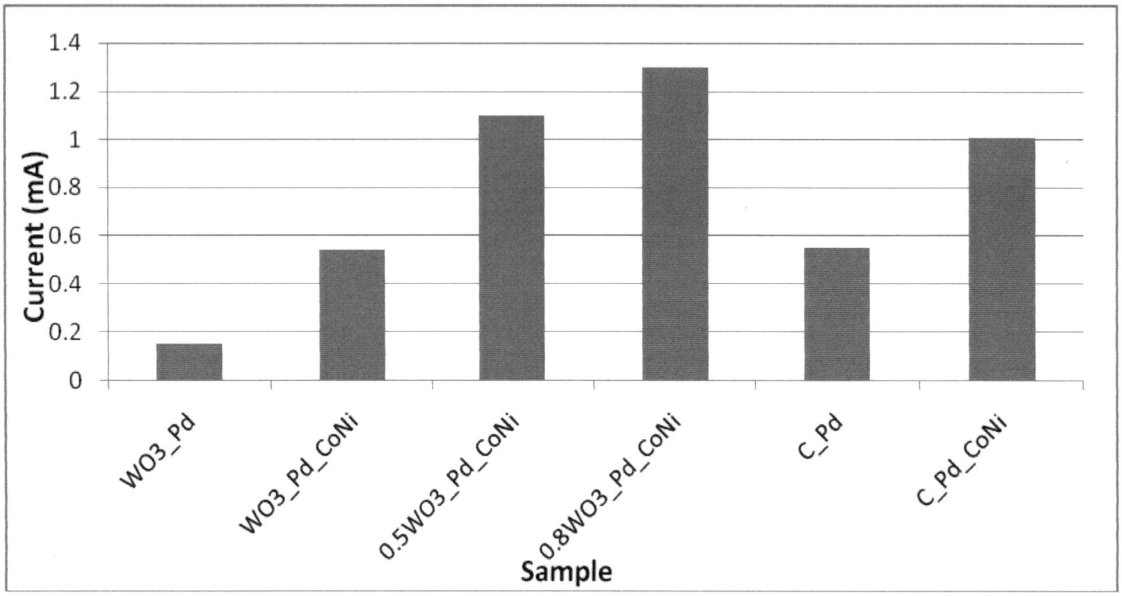

Figure 2. Peak Current of samples plated in the presence and absence CoNi at a scan rate of 25mV/s.

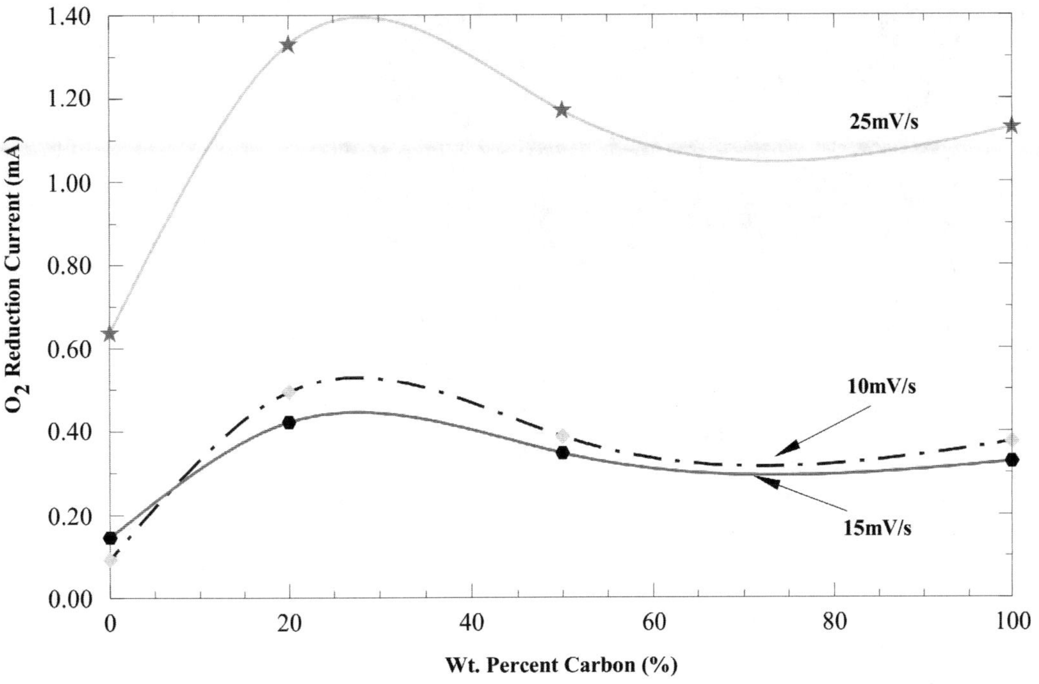

Figure 3. Oxygen reduction current on CoNi catalyst as a function of WO_3/C composition

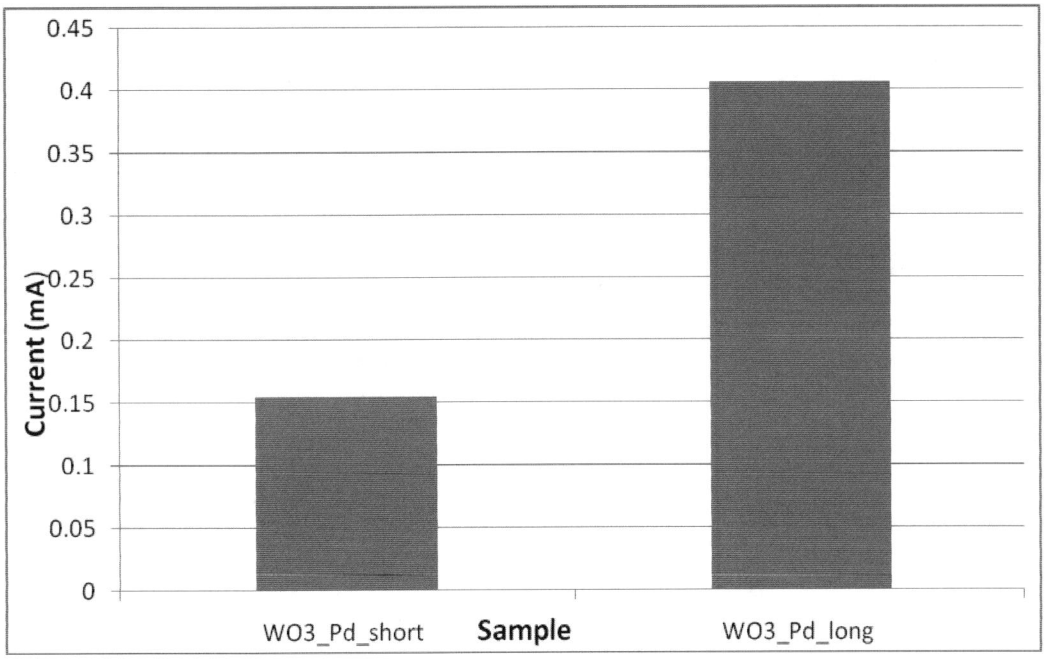

Figure 4. A comparison of Pd catalyzed WO_3 annealed for 24 hours and for one week at a scan rate of 25mV/s

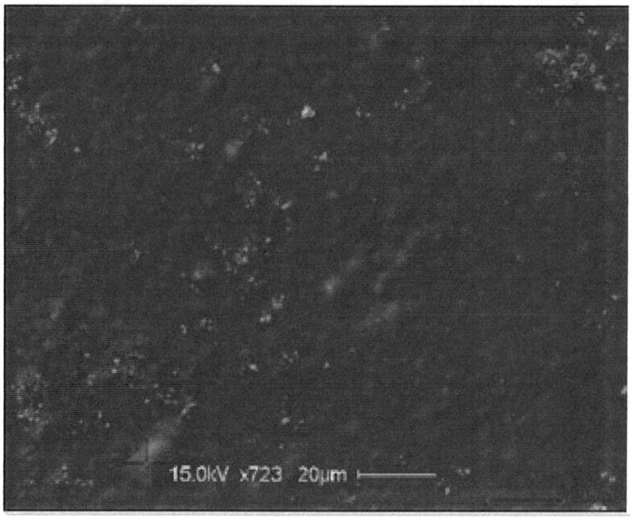

Figure 5a. EDAX results showing surface morphology of 50% WO_3/ 50% C plated with CoNi for 5 minutes

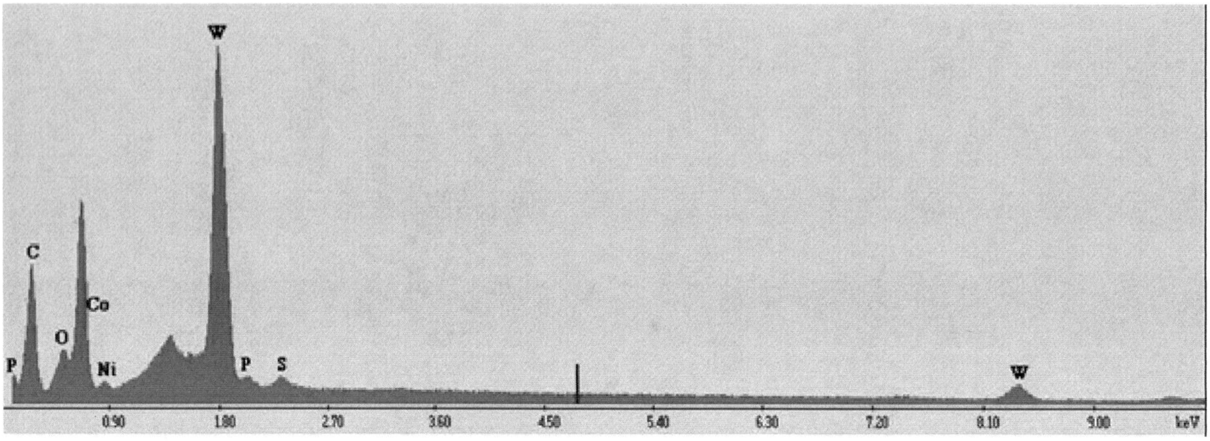

Figure 5b. Spectrum of selected area on the surface of the 50% WO_3/50%C substrate.

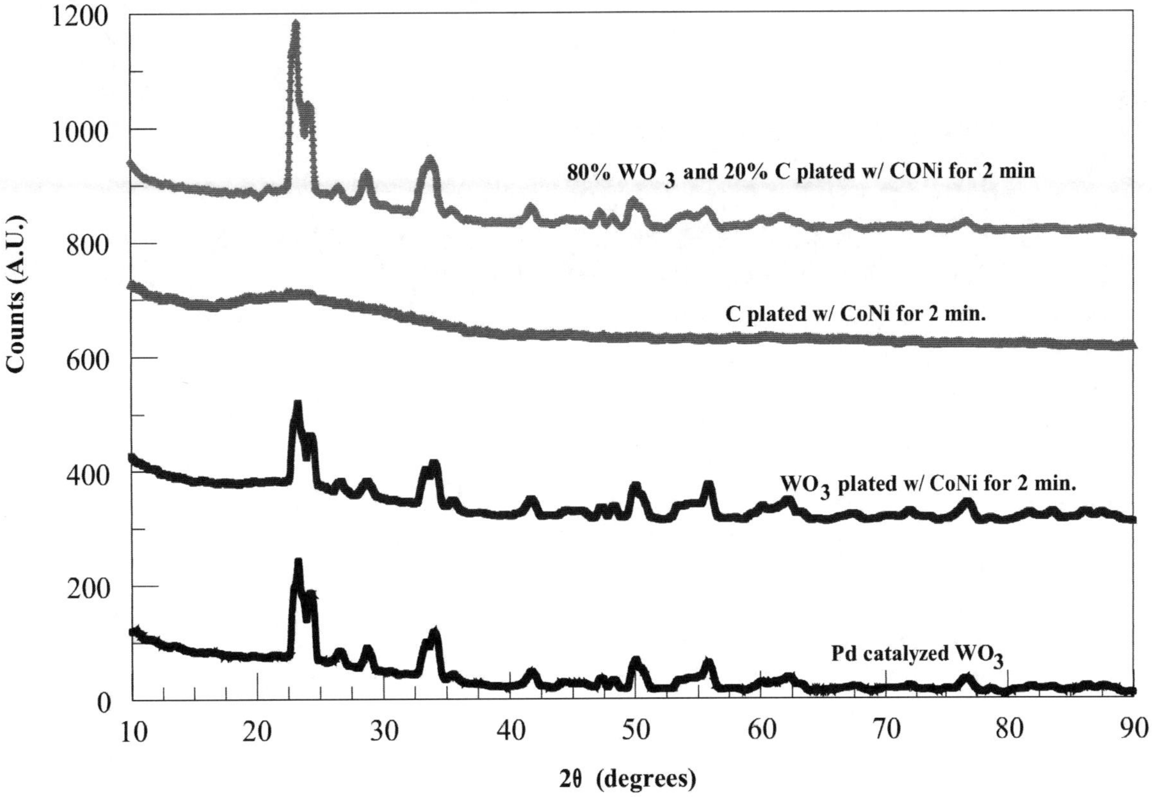

Figure 6. XRD results of WO_3 and C both plated with CoNi for 2 minutes and the results of Pd catatalyzed WO_3

Acknowledgments

We acknowledge the support of the Partnership for Research and Education in Materials initiative of the National Science Foundation under NSF Award 0351770, NSF Grant No. DMR-0351770.

References

1. Kangasniemi, K. H., D. A. Condit and T. D. Jarvi (2004), "Characterization of Vulcan Electrochemically Oxidized Under Simulated PEM Fuel Cell Conditions",*J. Electrochem. Soc.*, **151**, pp. E125 - E132
2. Willsau, J and J. Heitbaum (1984), "The Influence of Pt-activation on the Corrosion of Carbon in Gas Diffusion Electrodes – A DEMS Study", *J. Electroanal. Chem.*, **161**, pp. 93 – 101
3. Chhina, H, Campbell, S and Kesler, O (2008), "High surface area synthesis, electrochemical activity, and stability of tungsten carbide supported Pt during oxygen reduction in proton exchange membrane fuel cells", *J. Power Sources*, **179**, pp. 50 – 59
4. Chhina, H, Campbell, S and Kesler, O (2007), "Ex situ evaluation of tungsten oxide as a catalyst support for PEMFCs ", *J. Electrochem. Soc.*, **153**, pp. B533 - B539
5. Trogadas, P and Ramani, V (2008), "Pt/C-WO3 electrocatalysts for degradation mitigation in polymer electrolyte fuel cells" , *J. Electrochem. Soc.,* **155**, pp. B696 - B703
6. Shim, J, Lee, C. R, Lee, H. K, Lee, J. S and Cairns, E. J (2001), "Electrochemical characteristics of Pt-WO3/C and Pt-TiO2/C electrocatalysts in a polymer electrolyte fuel cell", *J. Power Sources*, **102**, pp. 172 – 177
7. Ferreira, P. J, La O, G. J, Shao-Horn, Y, Morgan, D, Makharia, R, Kocha, S and Gasteiger, H. A. (2005), "Instability of Pt/C Electrocatalysts in Proton Exchange Membrane Fuel Cells: A mechanistic Investigation", *J. Electrochem. Soc.*, **152**, A2256 – A2271
8. Bruijn, F. A, Dam, V. A. T, and Janssen, G. J. M (2008), " Review: Durability and Degradation Issues of PEM Fuel Cell Components", **8**, pp. 3 – 22
9. Yu, X and Ye, S (2007), "Recent Advances in Activity and Durability Enhancement of Pt/C Catalytic Cathode in PEMFC: Part II: Degradation mechanism and durability enhancement of carbon supported platinum catalyst", *J. Power Sources*, **172**, pp. 145 - 154

Effect of Hydrophobic Polymer Treatments on the Capillary Properties of Gas Diffusion Layers for Polymer Electrolyte Membrane Fuel Cells

Jeff T. Gostick, Marios A. Ioannidis, Michael W. Fowler and Mark D. Pritzker
Dept. of Chemical Engineering, University of Waterloo, Waterloo ON, N2L 5N5

1. Introduction

Increasing the current density of polymer electrolyte membrane fuel cells (PEMFC) leads to reduced stack size and materials cost, both of which are necessary if fuel cells are to become commercialized. Increasing current density, however, also means increased production of water inside the cell. The presence of liquid water in the porous electrode layers, particularly the gas diffusion layer (GDL), dramatically reduces the flux of gaseous reactants to the catalyst sites by blocking pores, thereby imposing mass transfer limitations on the maximum current that can be attained in a cell (1). To mitigate the negative effect of liquid water, a hydrophobic coating is usually applied to the GDL to alter its capillary properties. Although this approach is known to improve cell performance under high current density operation, the actual change in the GDL capillary properties is unclear since reliable experiments have not been available. In this paper, a recently developed method for measuring water-air-GDL capillary pressure curves (2) is applied to GDLs with varying amounts of PTFE coatings to quantitatively determine changes in water behavior. Access to such direct information about the capillary properties of GDLs will allow coating procedures to optimized and tailored to provide the desired water behavior in the cell.

2. Literature Review

Previous attempts to determine the air-water capillary properties of GDLs have not been entirely satisfactory. Although mercury intrusion porosimetry is a well established technique for measuring capillary properties of many porous materials, this method is not useful for GDLs for several reasons. Firstly, mercury is highly non-wetting to both the graphite substrate and the PTFE coating in GDLs, so it is insensitive to changes in the surface heterogeneity, which is precisely the information of interest. Secondly, converting mercury intrusion pressure data to an equivalent air-water pressure requires knowing the contact angle of mercury and water on the GDL surfaces. Even if a single contact angle can be determined for mercury in the GDL, which is not straightforward (3), the water contact angle will vary for the graphite and PTFE surfaces, making this conversion impossible without additional knowledge of the PTFE distribution. Furthermore, this conversion requires the use of the Young-Laplace equation based on a bundle-of-tubes model, which is not necessarily valid for highly porous and fibrous GDLs. Clearly, there is a need for direct measurement of the air-water capillary properties of GDLs.

In an earlier work (3) we reported on the method of standard porosimetry (MSP) for measuring capillary pressure properties with water as the working fluid. This technique was limited, however, since it could only measure along a path of decreasing water saturation and only for $P_C < 0$, where $P_C = P_L - P_G$. These early results suggested that GDLs possessed a network of hydrophilic pores since suction ($P_C < 0$) was required to remove water from the GDL. This finding was in contrast to the results of Benziger *et al.* (4) who found that positive liquid pressure was required for water to penetrate a GDL, even without a PTFE coating. Fairweather *et al.* (5) have recently reported a technique that uses a syringe pump to inject microliter volumes of water into a GDL. This technique, called the microfluidic approach (MFA), can scan along both increasing and decreasing water saturation paths and over a wide range of capillary pressures, $P_{C,MIN} < 0 < P_{C,MAX}$. Their results revealed for the first time a strong hysteresis effect, with water injection occurring at $P_C > 0$, as observed by Benziger *et al.* (4) and water withdrawal occurring at $P_C < 0$, as observed with MSP (3). Despite the insights gained by the MFA approach, this technique is not entirely satisfactory since the controlled parameter is the injected volume. The danger with adding volume in fixed amounts is that the pore space made accessible, V, at any given pressure, P_i, may not be completely filled by the amount of fluid injected, V_i (i.e. $V_i < V(P_i)$). A capillary pressure curve is generally expected to represent the amount of pore volume accessible at a given pressure, making the results of the MFA experiment difficult to interpret.

An alternative approach to measuring the water-air capillary pressure curves for GDLs has been recently reported (2). In this method the capillary pressure is controlled by adjusting the gas pressure above the sample while maintaining the liquid pressure at a constant value. Liquid enters the sample from a water reservoir and liquid saturation is monitored by tracking the changing mass of the reservoir with an analytical balance. Like the method of Fairweather et al. (5), this approach allows scanning along both increasing and decreasing saturation paths, and over a wide range of capillary pressures ($P_{C,MIN} < 0 < P_{C,MAX}$).

3. Experimental

In this work, the effect of PTFE loading on the capillary pressure curve is analyzed for two different substrate materials, Toray090 and SGL10. The specifications of each sample are given in Table 1. The PTFE loading values were specified by the supplier (SGL Carbon for the SGL10 samples and E-Tek for the Toray090 samples). The thickness of each sample was measured with a micrometer and porosity was determined from mercury intrusion porosimetry tests. The present set of experiments uses the setup and procedure detailed previously (2).

Table 1 – Specifications of samples tested

	Toray 090			SGL10		
Sample Name	090A	090C	090D	10AA	10BA	10EA
PTFE Loading [wt%]	0	10	20	0	5	30
Thickness [μm]	290	290	290	380	360	380
Porosity [-/-]	0.78	0.75	0.73	0.90	0.88	0.84

4. Results and Discussion

The results for the Toray 090 samples are given in Figure 1 and for the SGL10 samples in Figure 2. These curves show the results from -15000 to +15000 Pa to highlight the important features, although data were obtained for a range of -25000 to +30000 Pa for all samples. Each sample was tested from an initially dry state beginning at -15000 Pa, and then the capillary pressure was increased stepwise to 30000 Pa. The system was held at each pressure point until the saturation stopped changing. After 30000 Pa was reached the capillary pressure was decreased incrementally to a value of -25000 Pa. This loop from -25000 to 30000 and back to -25000 was repeated again. In general, the water injection followed a different path on the second injection, while the initial and secondary water withdrawal was identical. Further loops all followed the secondary injection path exactly and are not shown for clarity.

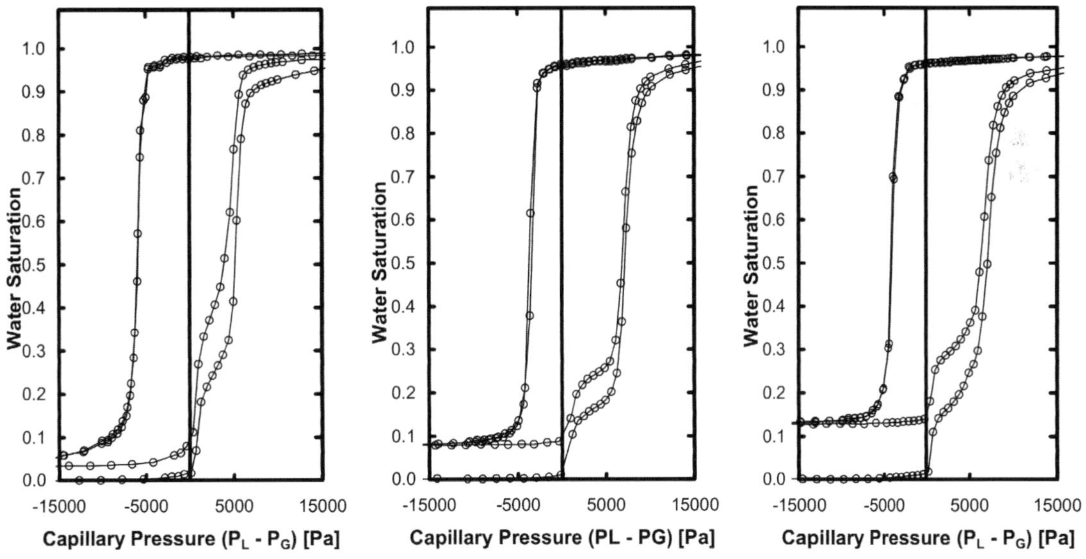

Figure 1 – Water-air capillary pressure curves of Toray 090. Left: 090A (0%PTFE). Middle: 090C (10%PTFE). Right: 090D (20%PTFE).

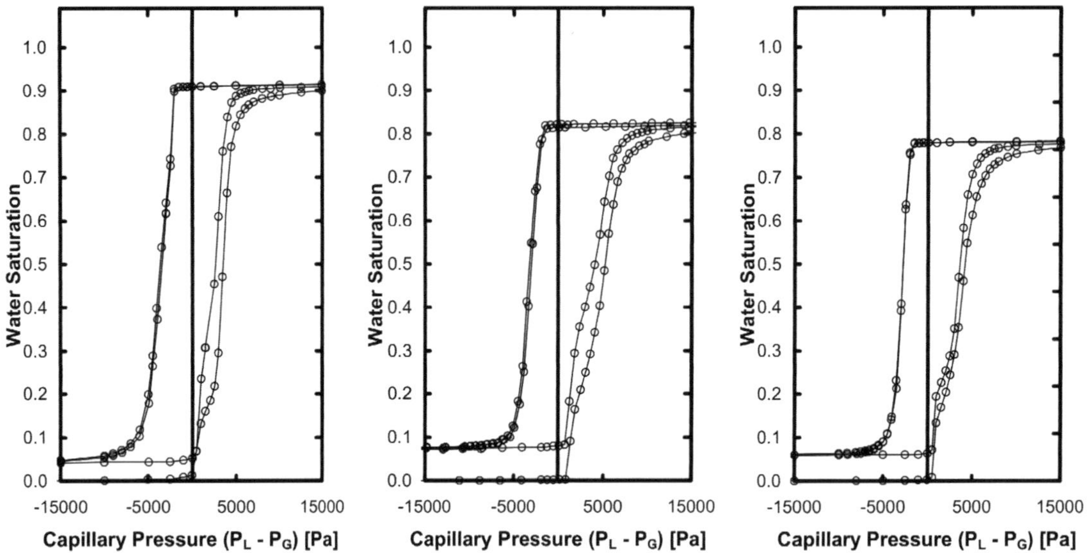

Figure 2 – Water-air capillary pressure curves of SGL10. Left: 10AA (0%PTFE). Middle: 10BA (5%PTFE). Right: 10EA (30%PTFE).

Perhaps the most striking feature of these curves is how little difference there is between different PTFE loadings on a given substrate. Intuitively, one might have expected the addition of PTFE to dramatically alter the shape of these curves. Instead, the differences are more subtle and quantitative. For instance, the water withdrawal for the 090A sample occurs at -4600 Pa while for the 090C and 090D samples it occurs at -2100 Pa. Similarly, water removal begins at -2100 for 10AA and at -1500 for 10BA and 10EA. This suggests that the addition of PTFE does significantly reduce the capillary pressure required to remove water, despite no apparent change in the qualitative features of the capillary pressure curves. Similar evidence can be found by comparing the primary injection legs of Toray 090 samples. The main injection (the sharp rise beginning at about s = 0.3) occurs at 4400 Pa for 090A and 6000 Pa for 090C and 090D. This indicates that the addition of PTFE increases the resistance to water penetration, as expected.

Another interesting feature of these results is that there appears to be little advantage to adding excessive PTFE. There are quantitatively noticeable differences between SGL10AA and SGL10BA or Toray090A and Toray090C, but there is no significant change between SGL10BA and SGL10EA or Toray090C and Toray090D. This is an important finding since it means that PTFE loading can be kept to a minimum, which enhances electrical conductivity of the GDL. This also suggests that higher PTFE contents result in thicker hydrophobic coatings, and not necessarily more hydrophobic surfaces, indicating that PTFE application procedures might be improved by finding ways of promoting more PTFE surface coverage.

5. Conclusions

The water-air capillary pressure properties of two gas diffusion layer materials for polymer electrolyte fuel cells were tested. The two substrate materials (SGL10 and Toray090) were each tested with varying amounts of hydrophobic polymer content. The addition of PTFE did not quantitatively alter the capillary pressure curve when compared to sample with no PTFE added. It was found, however, that significant quantitative differences, consistent with increased hydrophobicity, appeared when PTFE was added. The addition of excess PTFE did not noticeably alter the capillary properties from those of a modest PTFE loading. This indicates that excess PTFE creates thicker hydrophobic layers, but not more hydrophobic surface, and that an improved PTFE application technique could be sought to maximize the coverage of GDL internal surfaces.

6. References

1. Gostick JT, Ioannidis MA, Fowler MW, Pritzker MD 2007 Pore network modeling of fibrous gas diffusion layers for polymer electrolyte membrane fuel cells. J Power Sources 173:277-290

2. Gostick JT, Ioannidis MA, Fowler MW, Pritzker MD 2008 Direct measurement of the capillary pressure characteristics of water-air-gas diffusion layer systems for PEM fuel cells. Electrochem Comm In Press

3. Gostick JT, Fowler MW, Ioannidis MA, Pritzker MD, Volfkovich YM, Sakars A 2006 Capillary pressure and hydrophilic porosity in gas diffusion layers for polymer electrolyte fuel cells. J Power Sources 156:375-387

4. Benziger J, Nehlsen J, Blackwell D, Brennan T, Itescu J 2005 Water flow in the gas diffusion layer of PEM fuel cells. J Membrane Sci 261:98-106

5. Fairweather JD, Cheung P, St Pierre J, Schwartz DT 2007 A microfluidic approach for measuring capillary pressure in PEMFC gas diffusion layers. Electrochem Comm 9:2340-2345

768b Effects of Solvent-Casting Conditions on the Morphology and Properties of Disulfonated Poly(arylene ether sulfone) Copolymer Films for Polymer Electrolyte Membranes

Myoungbae Lee[1], Ozma Lane[2], Hae-Seung Lee[2], James E. McGrath[2], and Donald G. Baird[1]. (1) Chemical Engineering, Virginia Tech, Blacksburg, VA 24061, (2) Macromolecules and Interfaces Institute, Virginia Tech, Blacksburg, VA 24061

Disulfonated random and segmented multi-block copolymers based on poly(arylene ether sulfone) have been shown to be promising for fuel cell proton exchange membranes via a lab-scale batch solution casting followed by a mild drying. This batch casting approach is not appropriate for commercial production mainly because of its low drying rate. Thus, the effects of solution-casting conditions such as drying temperature, solvent type, etc. on the final morphology and properties were investigated for commercial production using a homemade batch convection drying apparatus installed to generate industrial-level drying conditions in the lab. In order to see the effect of molecular structure as well as casting conditions, a random copolymer and two different multi-block copolymers with different sequence lengths (5k and 15k) were prepared.

Despite a similar amount of proton-conducting moiety, the longer the block length was, the higher proton conductivity was achieved owing to more phase separation between sulfonated and non-sulfonated segments and subsequent well-defined continuous microstructure. No considerable effects of casting conditions on the microstructure and properties of the random copolymer were observed, whereas noticeable variations of the multi-block copolymers were induced; for example, the proton conductivity and water uptake of the multi-block copolymer films cast with a selective solvent (dimethylacetamide) were substantially higher than those of the equivalent films cast with a non-selective solvent (N-methylpyrrolidone). It was also observed that the multi-block copolymer with longer sequence length (15k) was more sensitive to the casting conditions. Our preliminary study also showed that the final morphology and properties also depended on substrate type, rate of solvent removal, and so on. Further investigation will be carried out in order to clarify the interrelationship among casting conditions, microstructures and properties. Also, this study will be extended to the development of measurement scheme of rheological properties (G' and G") of copolymers in solution-cast films during drying process for the study of transformation and kinetics of the copolymers in solution-cast films.

768c Electrochemical Performance of Direct Carbon Fuel Cell Based on Molten Carbonate Eutectic Electrolytes

Qinghua Liu[1], Levi T. Thompson*[2], Ye Tian[3], Chun Xia[3], Lijun Jia[3], and Yongdan Li**[3]. (1) Chemical Engineering, University of Michigan, 3200 H.H. Dow Building, 2300 Hayward St., Ann Arbor, MI 48109-2136, (2) Chemicall engineering, University of Michigan, 3200 H.H. Dow Building, 2300 Hayward St., Ann Arbor, MI 48109-2136, (3) Tianjin Key Laboratory of Applied Catalysis Science and Technology, Tianjin University, 92 Weijin Road, Tianjin, 300072, China

Direct carbon fuel cell (DCFC) has received much attention over the past decade. It represents a way to efficiently convert the chemical energy in solid carbon fuels (mainly from fossil fuels) directly into electrical energy and produce high concentration carbon dioxide that can be easily captured and sequestered. In this process, the thermal efficiency is as high as 100% because of the almost zero entropy change. Molten carbonate is employed as one kind of candidate electrolyte in DCFC in previous study. In this work, the electrochemical performance of a DCFC single cell being used three kinds of molten carbonate eutectic electrolyte with different composition was investigated. Anode fuel: graphite rod, diameter = 6.0 mm; Cathode: closed-bottom porous nickel tube, inner diameter = 41.6 mm, thickness = 4.0 mm, a mixture of CO2 and O2 (with the mol ratio of 2:1) was served as reactant in the cathode; Electrolyte: three kinds of carbonate melt were used, the property of them are listed in table 1; Operating temperature: 923.15 K. The electrochemical performance was measured by Computerized Electrochemical Instruments CBI660B (Chenhua Ltd. Shanghai). Fig. 1 shows the electrochemical performance of the single DCFC cell based on three kinds of molten carbonate electrolytes. The performance of 52.0Li/48.0Na electrolyte is as same as that of 62.0Li/38.0K electrolyte, and both are better than that of 43.5Li/31.5Na/25.0K electrolyte at low current density. However, at high current density, 52.0Li/48.0Na electrolyte performances best among them. The peak power densities are 10.29 mW cm-2, 22.65 mW cm-2, 25.66 mW cm-2 in the order of 43.5Li/31.5Na/25.0K, 62.0Li/38.0K, 52.0Li/48.0Na carbonate electrolyte. In conclusion, molten carbonate is a kind of promising electrolyte in DCFC. For various carbonates with different composition, 52.0Li/48.0Na carbonate melt has the best electrochemical performance. Key words: direct carbon fuel cell; molten carbonates; carbon
*Corresponding author, Tel.:+1-734-936-2015, Fax: +1-734-763-0459, E-mail: ltt@umich.edu (Levi T. Thompson) **Corresponding author, Tel.:+86-22-27405613, Fax: +86-22-27405624, E-mail: ydli@tju.edu.cn (Y.D. Li)

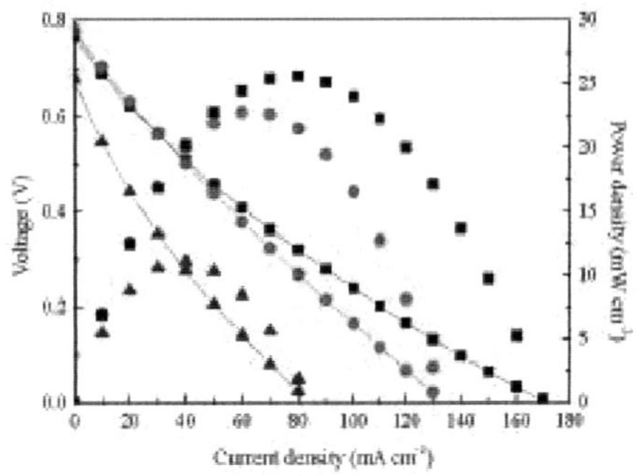

Fig. 1. Voltage-current density curves and power density-current density curves at 923.15 K. (-▲-): 43.5Li/31.5Na/25.0K, (-●-): 62.0Li/38.0K, (-■-): 52.0Li/48.0Na.

Web Page: www.engin.umich.edu/dept/cheme/people/thompson.html and http://ydli.net/

DETERMINING KINETIC AND MASS TRANSFER LIMITING BEHAVIOR OF A SOLID OXIDE FUEL CELL VIA AC IMPEDANCE

*Robert U. Payne, Ying Zhu, Wenhua H. Zhu, and Bruce J. Tatarchuk**
Center for Microfibrous Materials Manufacturing
Department of Chemical Engineering
207 Ross Hall, Auburn University, AL 36849

S. Elangovan and Mark Timper
Ceramatec, Inc.
2425 South 900 West
Salt Lake City, UT 84119

Abstract

AC Impedance has been applied to solid oxide fuel cells at various operating conditions. Experiments have been performed on small (2 cm^2) button cells and larger (61 cm^2) planar cells. Button cells were fabricated with anode and cathode reference electrodes to facilitate separate testing of anode and cathode. Impedance spectroscopy has also been applied to a five cell planar stack and to single cells therein. Equivalent circuit models are derived with each circuit element representing different processes that limit cell output. Values for the elements are obtained by fitting the model to the experimental data. The degree to which electrode kinetics and mass transfer inhibit cell output may be evaluated from the fitted resistances, capacitances, and inductances of the circuit elements. The model (including the elements used, their arrangement, and the processes to which they are ascribed) is validated by changing cell temperature, reaction rate, and reactant concentration. The presented work is useful for performing stack maintenance and diagnostics and for designing process controls and power conditioning circuitry.

Introduction

Impedance spectroscopy is a novel and powerful method to investigate the dynamic response characteristics of an electrochemical system. AC impedance is performed by superimposing an AC signal on the DC output of an electrochemical cell and measuring the impedance over a spectrum of frequencies. Equivalent circuit models, circuits of electrical elements producing a similar load response to the device under investigation, are derived from experimental data. Equivalent circuits are beneficial for stack diagnostics, because the different circuit elements theoretically represent different relaxation processes occurring in the electrochemical cell(s); hence, fit data can be used to distinguish which processes are limiting cell behavior and to what extent. In the following work, AC impedance spectroscopy has been applied to test cells and stacks at various operating conditions, and fitted equivalent circuits have been derived.

* Corresponding author. Tel.: +1-334-844-2023. E-mail address: brucet@eng.auburn.edu

Experimental

AC impedance was applied to solid oxide fuel cells (Ceramatec, Inc.) of two different sizes: small (2 cm^2) button cells and larger (61 cm^2) planar cells. Both consisted of a zirconia based electrolyte (~170 μm thick), a nickel based anode, and a manganite based cathode. The cells were fed at ~15 psig with air on the cathode side and mixtures of hydrogen and nitrogen bubbled through water on the anode side while varying load, temperature, reaction rate and reactant concentration. Small reference electrodes bonded to both sides of the button cell facilitated separate testing of the anode and cathode. The planar stack was fed with 1.5 SLPM air and 2.2 SLPM differing anode mixtures of hydrogen and nitrogen bubbled through water. Data were acquired at different stack currents for the cells on both ends of the stack (cell #s 1 and 5) and the middle cell (cell #3), as well as for the entire stack. All spectra collected from either cell size were fitted to equivalent circuits with Gamry's Echem Analyst software using the Levenberg-Marquardt algorithm.

Results and Discussion

Based on assumed reaction mechanisms, impedance data were analyzed in terms of equivalent circuits. Anode and cathode data of the button cell were fitted to their respective circuits to determine how reaction kinetics and mass transfer inhibits cell output. Total-cell equivalent circuits were then based on a combination of the anode and cathode models. The button cell analyses were then applied to the planar cells.

Button Cell

The overall anode reaction in an SOFC fed with hydrogen is the hydrogen oxidation reaction (HOR):

$$H_2 + O_O^{\times} \leftrightarrow H_2O + V_O^{\bullet\bullet} + 2e' \qquad (1)$$

(in Kröger-Vink notation)

Because several mechanisms [1, 2, 3, 4] are possible for the HOR, the equivalent circuit model [5], shown in Figure 1, is chosen based on the experimental data.

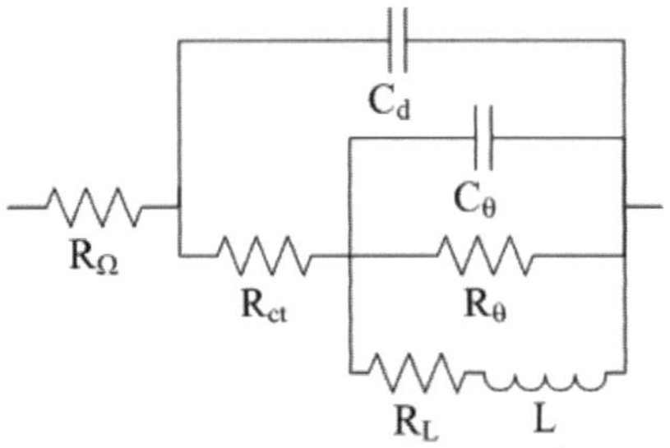

Figure 1. Equivalent circuit model for SOFC anode.

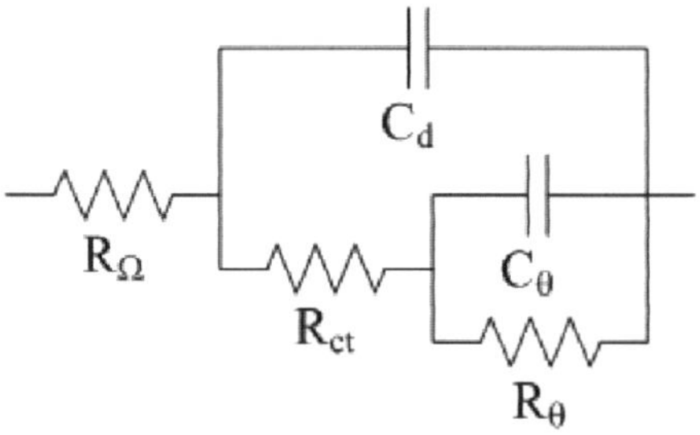

Figure 2. Equivalent circuit model for SOFC cathode.

The half reaction occurring at the cathode is the oxygen reduction reaction (ORR). Assuming a simple reaction mechanism [6, 7] of

$$O_2 + 2M \leftrightarrow 2MO \tag{2}$$

$$MO + 2e' + V_O^{\bullet\bullet} \leftrightarrow M + O_O^{\times} \tag{3}$$

(both in Kröger-Vink notation) and negligible mass transfer, a common equivalent circuit for reactions consisting of one adsorbed intermediate [8] may be derived (Figure 2); however, a CPE is expected here to replace C_d because of the variety of surfaces on which oxygen can react.

Adding the anode equivalent circuit in series with that of the cathode, a full cell equivalent circuit would theoretically be derived; however, an extra low frequency loop existed that is unique to the total-cell data. A finite diffusion element (FDE) was added to a simplified combination of the anode and cathode models (Figure 3) to fit the low frequency loop, because low frequency impedance is often attributed to concentration losses.

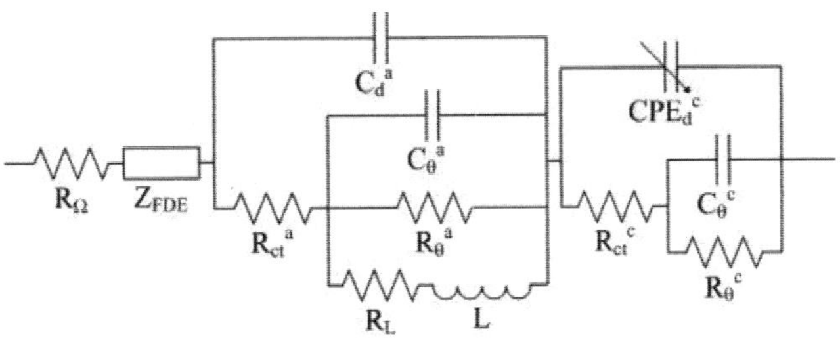

Figure 3. Equivalent circuit model for SOFC total-cell.

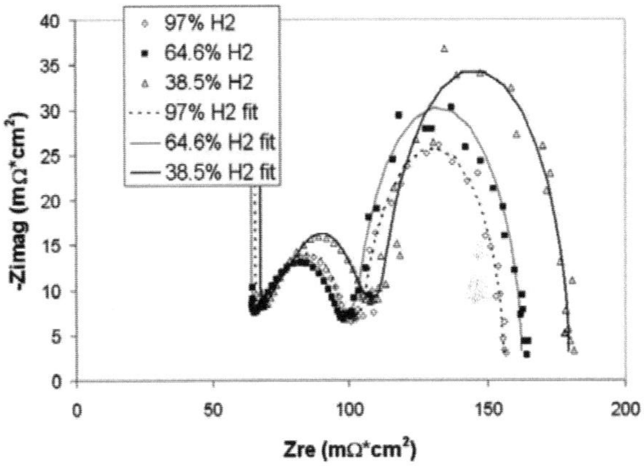

Figure 4. Prismatic cell entire stack data fit.

Prismatic cell

Theoretically, the impedance spectra of the planar cell stack would appear similarly in shape to that of the total button cell at similar operating conditions but scaled to a larger size; however, non-idealities inevitably exist as electrode areas are increased. Due to the sheer size of the cells, the current, temperature, and reactant concentration are distributed along the active surface. Additionally, the thinner bonding layers of the full cell decrease the diffusion layer thickness.

The button cell analyses were applied to the individual cells and then the entire fuel cells. The arrangement in Figure 3 was well fitted to the entire stack data (Figure 4).

Acknowledgement

The authors would like to thank Dennis Larsen, Lyman Frost, Anthony Nickens and everyone else at Ceramatec Inc. who helped make this project possible.

References

1. A. Bieberle, L. J. Gauckler, Solid State Ionics, 146 (2002) 23–41.
2. Wolfgang G. Bessler, J. Electrochem. Soc., 153 (2006) A1492–A1504.
3. Huayang Zhu, Robert J. Kee, Vinod M. Janardhanan, Olaf Deutschmann, and David G. Goodwin, J. Electrochem. Soc., 152 (2005) A2427–A2440.
4. V. N. Chebotin, M. V. Glumov, A. D. Neuimin, S. F. Palguev, Sov. Electrochem. Soc., 7 (1971) 55–61.
5. D. B. Zhou, H. Vander Poorten, J. Electrochem. Soc., 145 (1998) 936–945.
6. A. Lasia, in: B. E. Conway, J. O'M. Bockris, R. E. White (Eds.), Mod. Asp. Electrochem., 32, Kluwer Academic / Plenum Publishers, New York, 1999, pp. 143–242.
7. J. Deseure, Y. Bultel, L. Dessemond, E. Siebert, P. Ozil, J Applied Electrochem., 37 (2007) 129–136.
8. A. Bieberle, L. P. Meier, and L. J. Gauckler, J. Electrochem. Soc., 148 (2001) A646–A656.

768e Nonlinear Analysis of Rate Controlling Mechanisms of Solid Oxide Fuel Cell Cathodes

Cortney Kreller and Stuart Adler. Chemical Engineering, University of Washington, Box 351750, Seattle, WA 98195

Most potential losses in intermediate temperature (500-700 C) solid oxide fuel cells stem from the cathode. To engineer better SOFC cathodes, the processes limiting performance must be identified and linked to specific material properties and morphological features. Transition metal perovskite oxides, such as $La_xSr_{1-x}CoO_{3-\delta}$ are a promising class of materials that possess good electronic and ionic conductivity and catalyze the oxygen reduction reaction in the target temperature range.

In order to gain insight on the physical processes governing cell performance, we have used nonlinear electrochemical impedance spectroscopy (NLEIS) to probe the nonlinear harmonic current-voltage characteristics of a $La_xSr_{1-x}CoO_{3-\delta}$ electrode. Harmonic spectra acquired over a range of conditions were compared to a theoretical model that considers the relative roles of material thermodynamics, surface exchange reaction and ion diffusion within a cylindrical electrode particle. For a Strontium dopant content of 0.4, a comparison of the theoretical and experimental nonlinearities suggests that the cell behavior is not consistent with a purely bulk-diffusion limited mechanism. This result is contrary to linear impedance observations and suggests that the nonlinear behavior could be influenced by an alternative surface diffusion mechanism or material thermodynamics.

768f Evidence for Hydrogen Spillover at Extremely Low Pressure for Pt- and Pd- Supported on Carbon Materials

Angela D. Lueking and Qixiu Li. Energy and Mineral Engineering, The Penn State University, 120 Hosler Bldg, University Park, State College, PA 16802

The mechanism of hydrogen spillover has long been explained by a chemical kinetics analysis, which leads to a P^0.5 pressure dependence of the spillover process, suggesting the uptake will be appreciably increased as pressures are extrapolated to those of interest for DOE hydrogen storage targets (i.e. 100 bar). However, using a high-pressure gravimetric balance, we have experimentally observed mass uptake that can only be attributed to hydrogen spillover at extremely low pressures (i.e. 10-7 bar), inconsistent with models that predict a P^0.5 pressure dependence. The uptake is observed for samples containing metal and carbon, and not observed for inert samples or carbon samples without metal. Our results also indicate that pretreatment is a key parameter to observe the low pressure uptake. The results have important implications as to how overall uptake is calculated in gravimetric versus volumetric adsorption measurements: for the latter case, the low-pressure spillover is not detectable. Mechanistic implications of hydrogen spillover for hydrogen storage will be discussed.

AUTHOR INDEX

AUTHOR INDEX

American Institute of Chemical Engineers
3 Park Avenue
New York, New York 10016-5991

CURRAN ASSOCIATES INC
proceedings.com